Lecture Notes in Mathematics

A collection of informal reports and seminars
Edited by A. Dold, Heidelberg and B. Eckmann, Zürich

312

Symposium on Ordinary Differential Equations

Minneapolis, Minnesota, May 29–30, 1972

Edited by William A. Harris, Jr.
University of Southern California, Los Angeles, CA/USA
and Yasutaka Sibuya
University of Minnesota, Minneapolis, MN/USA

Springer-Verlag
Berlin · Heidelberg · New York 1973

AMS Subject Classifications (1970): 34-02, 34A20, 34A25, 34A50, 34B10, 34B15, 34C05, 34C25, 34E05, 34E15, 34J05, 45M10, 39-02, 39A15, 49-02, 49A10

ISBN 3-540-06146-0 Springer-Verlag Berlin · Heidelberg · New York
ISBN 0-387-06146-0 Springer-Verlag New York · Heidelberg · Berlin

Offsetdruck: Julius Beltz, Hemsbach/Bergstr.

This volume, as well as the Symposium itself, is dedicated to Hugh L. Turrittin for the many contributions he has made over the past years to the development of this subject. He has influenced most of us directly or indirectly through his inspiration, encourgement and guidance. In life as well as in Mathematics he has been a true friend and a true companion.

PREFACE

This volume is the proceedings of a SYMPOSIUM ON ORDINARY DIFFERENTIAL EQUATIONS that was held May 29-30, 1972 at the University of Minnesota, honoring Professor Hugh L. Turrittin upon his retirement, the theme of the SYMPOSIUM was current researches arising from central problems in differential equations with special emphasis in the areas to which Professor Turrittin has made contributions.

The first paper of these proceeding entitled "My Mathematical Expectation" by Professor Turrittin sets the tone and clearly establishes the scope and breath of the SYMPOSIUM.

The editors wish to thank Professor J. C. C. Nitsche, Head of the School of Mathematics, the Symposium Committee and members of the School of Mathematics for their support of the Symposium and their warm hospitality and generosity.

September 1972

William A. Harris, Jr.
Yasutaka Sibuya

CONTENTS

My Mathematical Expectations

H. L. Turrittin

1. Introduction

This expository presentation is intended to be a brief review of my mathematical research, beginning with my first paper [1], in 1936. Since I am about to retire, it is an appropriate occasion to look once again at my publications in a critical fashion.

This Symposium on Ordinary Differential Equations is, I feel, indeed a very particular and gratifying expression of esteem on the part of my colleagues and students. The papers presented here at the Symposium have been of special interest and have provided me with several mathematical inspirations which, it is hoped, will materialize later in the form of specific results. I wish to express my gratitude to all those who have taken part in the Symposium, to those who have helped organize it, and especially I wish to thank Professor Yasutaka Sibuya and Professor William A. Harris, Jr.

The title is intended to indicate that the emphasis will be placed on what I had expected to do as I initially attacked each new problem, rather than on what I actually accomplished. In most cases there is quite a difference in these two things, as I shall point out as we proceed.

As a retiring professor there is no point at this time in trying to conceal mathematical ideas. They are likely known many of you anyway. In short I am not trying to stake out claims to certain unsolved problems for future study. Precisely the contrary, since I did not succeed in solving certain problems, nothing would please me more mathematically than to see others find the solutions or extend or use some of my results.

2. Equations involving a parameter

In 1930 I became a graduate student in mathematics at the University of Wisconsin and Professor Rudolph E. Langer became my graduate adviser and mentor. Langer had already become interested in what he called turning point problems in the theory of ordinary differential equations, see [15-18]. Such problems may occur when the differential equation contains a parameter. At Professor Langer's suggestion for my Ph.D. thesis I studied the solutions of an equation of the type

$$\frac{d^n y}{dx^n} + \sum_{i=1}^{n} \rho^{ir} P_i(x, \rho) \frac{d^{n-i} y}{dx^{n-i}} = 0, \tag{1}$$

where $P_i(x, \rho) = \sum_{j=0}^{\infty} P_{ij}(x)/\rho^j$, $(i = 1, \ldots, n)$, is convergent if $a \le x \le b$ and $|\rho| > R > 0$. Here ρ is a large parameter. It was known in those days that, if the n roots $\phi_i(x)$, $(i = 1, \ldots, n)$, of the characteristic equation

$$\phi^n + P_{10}(x)\, \phi^{n-1} + P_{20}(x)\, \phi^{n-2} + \cdots + P_{n0}(x) = 0,$$

as functions of x were distinct on the interval $[a, b]$, then Equation (1) possessed n linearly independent formal asymptotic series solutions in the sense of H. Poincaré. These series solutions ran in powers of $1/\rho$. My problem was to extend these known results to the case in which the roots $\phi_i(x)$ might be multiple. This I succeeded in doing by using power series running in fractional power of ρ^{-1}.

To confess the truth in regard to this first paper, I did not have, quoting Dickins, "Great Expectations". Mine were quite simple at that time, get enough results to please Professor Langer and obtain the Ph.D. degree. Moreover on rereading the published version [1] of my thesis, it strikes me to-day as far too abbreviated. Hence, if a reader is interested in the topic under consideration, I recommend instead the examination of my paper [4] written sixteen years later.

This latter paper was written at the suggestion of Solomon Lefschetz. He had invited me to come to Princeton during my sabbatical leave and work essentially on the same problem as in my doctoral dissertation, but this time from a more general point of view and begin with essentially a system

$$\varepsilon^h \frac{dY}{dx} = \sum_{k=0}^{\infty} \varepsilon^k A_k(x) Y, \tag{2}$$

where the unknown Y is an n-component vector; $\varepsilon = \rho^{-1}$; and $a \leq x \leq b$. The basic hypothesis in this Princeton paper was that the Jordan classical canonical form of the given square matrix $A_0(x)$ would have the same form for all values of x on the interval $a \leq x \leq b$; i.e. as x varies, the size of the blocks and the number of 1's in each Jordan block would not change.

In regard to this Princeton paper [4], there were expectations. The hypothesis just described was made purposely to sidestep the entire turning point problem, i.e. the study of the behavior of solutions of Equation (2) when ε is small and x is in the neighborhood of one of those values for which there is a sudden change in the structure of the Jordan form of $A_0(x)$. My original expectation was to continue my Princeton work and to examine the turning point problem in great detail; but actually I never found the time to attempt to do this far more difficult piece of analysis. In fact my only results in this direction are contained in paper [7], a joint piece of work with William A. Harris, Jr.

Fortunately a number of mathematicians have continued the study of the turning point problem, although even to this day in particularly difficult cases, much remains to be done. For one interested in turning point problems, he may refer to the work of Y. Sibuya, [19, 20] and W. Wasow, [22,23], and particularly to [21], Chapter 8.

No attempt is made in this paper to give a complete bibliography relating to the topics here considered and, I hope, those not mentioned, who have made substantial contributions will forgive me for my omissions.

3. Zeros of exponential sums

With a Ph.D. degree in hand I was fortunate in 1934 to obtain a teaching position at El Paso, Texas, in the College of Mines and Metallurgy, for that year about one hundred Ph.D's in mathematics in the United States were without work, according to a report of the secretary of the American Mathematical Society. At El Paso, at the time, research was not stressed; but good teaching was much emphasized. Nevertheless I was looking for a new problem that I had some chance of solving.

L. A. Mac Coll, [24], had just published a paper giving asymptotic approximations for the number and location of the zeros of a function of the form

$$E(z) = \sum_{j=1}^{J} P_j \exp\{q_j(z)\},$$

where each $q_j(z)$ represents a polynomial in z and each P_j is a constant. Earlier R.E. Langer, [25], had studied the zero distribution when each P_j represents an analytic function behaving as a power of z, $|z|$ large, and each $q_j(z)$ is a polynomial of the first degree. I thought it would therefore not be too difficult to determine the asymptotic distribution of the zeros of $E(z)$ far from the origin when both the P_j and $q_j(z)$ were given polynomials of any specified degree in z. As one might judge from the date of publication of my results, see [2]. Only after learning something about Diophantine approximation and discovering some much needed facts in one of Van der Corput's papers, [26], was I able to get satisfactory results.

One has certain emotional reactions to one's published papers. I have always been pleased with this second paper, likely because I worked so hard and so long on it, and yet in a sense this paper turned out to be a disappointment, since in most applications no one has needed my analysis, either Langer's or Mac Coll's results sufficing for their purposes.

Furthermore there were certain expectations regarding paper [2] that did not materialize. The large zeros turned out to be located in certain bands extending out to infinity in the complex z-plane. Thinking in terms of polar coordinates, the radial approximate zero distribution in each band was determined; but I had hoped to have also something fairtly precise about the angular distribution of the zeros in each band; but on this latter problem no success.

Moreover a sequel to paper [2] was planned. I had expected to have a paper indicating a numerical procedure for evaluation of the greatest lower bound and the least upper bound of an expression

$$\left| \sum_{j=1}^{J} p_j e^{iQ_j(t)} \right| ,$$

where the p_j are given real constants, $i = \sqrt{-1}$, and the $Q_j(t)$'s are polynomials in t with real coefficients and t varies over the range $-\infty < t < \infty$; but here too I was unsuccessful.

4. Stokes multipliers

My third paper, [3], was concerned with the behavior in the large of solutions of the equation

$$(3) \qquad \frac{d^n w}{dz^n} - z^v w = 0 \quad , \qquad (v = 1, 2, \dots).$$

Such an equation has n linearly independent solutions $w_i(z)$, $(i = 1, \dots, n)$, which are entire functions; i.e. the $w_i(z)$ can be expressed as power series in z, convergent for all finite values of the complex variable z. Also the equation has n linearly independent solutions $\hat{w}_j(z)$, $(j = 1, \dots, n)$, which can be represented in appropriate sectors of the complex z-plane by formal asymptotic series running in fractional negative powers of z. If one wishes to determine the behavior of a particular solution $w_i(z)$ as $z \to \infty$ in some one of the appropriate sectors S_k, then it becomes necessary to evaluate the so-called Stokes multipliers c_{ijk} interrelating the solutions $w_i(z)$ and $\hat{w}_j(z)$ by means of the equation

$$w_i(z) = \sum_{j=1}^{n} c_{ijk} \hat{w}_j(z), \quad (i = 1, \ldots, n).$$

By using Barnes integral representations

$$w(z) = \int_C \varphi(s) z^{ms} ds$$

of solutions, where the constant m, path C in the complex s-plane, and function $\varphi(s)$ are judiciously chosen. I did succeed in finding a specific set of linear algebraic equations, which, when solved, would determine the desired Stokes multipliers. I left the problem at this stage, with unforeseen results to follow.

Sixteen years later I returned to Equation (3), expecting to get Stokes multipliers when v was a negative integer, and succeeding only for all n when v = - 1 and v = -2, because in working out paper [5] I had become well acquainted with the use of Laplace integrals, and foolishly decided to abandon Barnes integral representations of solutions and instead try Laplace integral representations of the form

$$w(z) = \int_C e^{zs} \varphi(s) ds.$$

In retrospect I now see my approach to Mathematics has been too narrow and specialized. My choice of a new problem was usually the generalization or extension of an old problem found in the literature. I should have also branched out and tried to create some entirely new phase of mathematics, for indeed, since 1936 many new types of mathematics have been created and developed.

My main technique of solving a new problem is likely becoming apparent. I always began by attacking the simplest special problem that I could think of involving one of the new features I hoped to introduce; then make the problem slightly harder by incorporating perhaps two new complications, and so on; thus hopefully and methodically to try to reach the final desired result, as I grew to understand the nature of the problem and its complications.

Further I was always handicapped by a lack of knowledge of modern algebra and topology. Although I realized this, I studied these topices unfortunately nowhere near thoroughly enough.

In regard to Equation (3) I had expectations, for I wanted to get the Stokes multipliers for all integer values of v and vaguely intended to then study the distribution of zeros of the solutions of such differential equations, based on my study of exponential sums, [2]. This I never did.

However J. B. McLeod in 1966, [27], used my results relating to Equation (3) in his study of the distribution of eigenvalues of a certain self-adjoint boundary value problem associated with the equation

$$\frac{d^{2n}y}{dx^{2n}} + (-1)^{n-1}(\lambda - x)\,y = 0$$

on the interval $0 \leq x < \infty$. Unfortunately, as I have indicated, I had not given the stokes multipliers explicitly, but only as a solution of a rather complicated system of linear algebraic equations. This prevented Mc Leod from doing quite as satisfactory a bit of analysis as he would have liked to do.

B. L. J. Braaksma noticed this and reattacked the problem going back to the use of Barnes integrals, see [28,29]. He obtained elegantly and explicitly the Stokes multipliers for Equation (3), not only in the case when v is an integer, but when v is any real or complex constant.

Braaksma's methods undoubtedly can be applied to certain other types of differential equations, such as those described by H. Scheffé in one of his papers, see [30]. Thus with Braaksma's work I am pleased to see a particular problem completed in a most satisfactory fashion with more applications sure to come.

5. Summation of asymptotic series solutions

From time to time I had been studying summation of divergent series, particularly by Cesaro and Borel summability. My hope and expectation was to replace by convergent series the usually divergent asymptotic series representing solutions of the matrix differential equation

$$(4) \qquad \tau^g \frac{dX}{d\tau} = \sum_{k=0}^{\infty} \tau^k A_k X,$$

where g is a positive integer, $X = X(\tau)$ is the unknown n by n square matrix solution to be found, and the A_k's are given n by n square matrices such that the indicated sum is convergent for all sufficiently small values of the complex independent variable τ. My paper [5] on this topic has become my most quoted and used piece of analysis.

For instance Nicolas M. Katz, at Princeton, in a seminar on degeneration of algebraic varieties in 1969-70 has made good use of my results. Also in connection with paper [5] there is presently considerably interest in determining necessary and sufficient conditions relating to Equation (4) that guarantee that the origin $\tau = 0$ is a regular singularity and, in the more general case, the determination of invariants measuring the amount of irregularity of the singular point $\tau = 0$, see the recent work of R. Gérard of Strasbourg and and A.H.M. Levelt of Nijmegen, [36, 37] and the references in these two papers.

The important part of paper [5] has not turned out to be the summation of the divergent asymptotic power series solutions, which in some cases were replaced by convergent generalized factorial series; but instead the chief interest seems to relate to the canonical form to which the original equation can be reduced. From my own point of view this paper [5] was not a complete success, for I was not able in all cases to sum the divergent series and replace them by convergent series. Thus I still have the expectation that some one will get an inspiration that will finish the job.

I also studied the problem of the summation of the divergent asymptotic power series solutions of a nonhomogeneous equation of type (4) in the form

$$\frac{dX}{ds} = \sum_{j=0}^{\infty} s^{-j} A_j X + \sum_{j=0}^{\infty} s^{-j} B_j$$

as $s \to \infty$ and again summed the formal series solutions in some cases and failed to do the summation in other cases.

6. Peculiar solutions of a nonlinear equation

In 1956, while on a joint University of Minnesota and Minneapolis-Honeywell research project, I became interested in the periodic solutions of a modified Duffing's equation of the form

$$(5) \qquad \frac{d^2x}{dt^2} + h\frac{dx}{dt} + x + x^3 = b\cos wt + [a^3 \cos 3\,wt]/4$$

where

$$b = a - aw^2 + 3a^3/4,$$

and $h, w,$ and a are real constants, and h is positive and small.

A periodic solution in general of period $2\pi/w$ would have the usual Fourier series representation

$$x(t) = [\frac{a_0}{2} + \sum_{n=1}^{\infty} (a_{2n} \cos 2nwt + b_{2n} \sin 2nwt)]$$

$$+ [\sum_{n=0}^{\infty} (a_{2n+1} \cos \{2n+1\}\, wt + b_{2n+1} \sin\{2n+1\}\, wt],$$

where the mixed harmonic solution $x(t)$ is split into its even harmonic and odd harmonic parts as indicated by the brackets. Dana Young and others on the project believed at first that, since the forcing term, represented by the right-hand member of Equation (5) was itself an odd harmonic function of time t, the response, that is the periodic solution, would necessarily itself be an odd harmonic function for all values of a and w and very small h and not a mixed harmonic. My problem and purpose, working jointly with W. J. A. Culmer, was to prove this is false.

We were able to show, see [6], that, if the point (a, w) were on the edge of certain narrow instability regions in the real (a, w) - plane, one would get an unexpected mixed harmonic response. An analogue computer indicated and we expected that one could show that, if the point (a, w) were not only on the edge, but also inside the narrow instability regions in the (a, w)-plane, the same peculiar phenomena would occur. We never succeeded in proving this.

Warren Loud was at the time on the same research project and he and some of his students have, I am pleased to report, pursued the study of nonlinear phenomena well beyond this special example; see [31-33].

7. Difference equations

In the late 50's I had become interested in ordinary linear difference equations, as well as differential equations. At first I studied the nature of the formal power series solutions of equations of the type

$$X(s+1) = s^g \sum_{k=0}^{\infty} s^{-k} A_k X(s),$$

where g is a positive integer and the A_k and $X(s)$ are n by n square matrices. A simplified canonical form for such equations was obtained. Two papers, [9, 10], were devoted to this topic, paralleling my work on differential equations. W. J. A. Culmer and W. A. Harris, Jr., [34], investigated summing the divergent asymptotic series solutions of difference equations and replacing them by convergent factorial series, only to find with the methods at their disposal that they could succeed in some, but not in all cases.

On becoming more familiar with difference equations and their close relation to differential equations, I was in hopes that the theory of difference equations could be brought completely abreast with that for ordinary differential equations. W.A. Harris, Jr., Y. Sibuya, S. Tanaka, and others have made very considerable progress in this direction, see the bibliography in [35] for references.

Here I had one grand expectation (undoubtedly showing my naiveté at the time), namely that one could produce an overall theorem essentially stating that all, or nearly all, theorems about ordinary differential equations were just limits of theorems about difference equations as the difference interval approached zero in length. But this is undoubtedly a dream, for what theorem in difference equations would produce in the limit the Poincaré-Bendixson theorem on limit cycles? Or even far more difficult, is there a theorem in difference equations which in the limit would make it clear whether or not there was an analogue of the Poincaré-Bendixson theorem for the three dimensional phase space when all trajectories enter a torus containing no critical points?

This brings me in the next two sections to two of my papers which George D. Birkhoff, had he lived to see them, might perhaps have called "cute little papers".

8. A real canonical Jordan form

Niemitski and Stepanov in the original Russian edition of their text on "Qualitative Theory of Differential Equations" were interested, among many other things, in the precise form of the real solutions of a vector differential equation

$$\frac{d\vec{x}}{dt} = A\vec{x}, \quad (-\infty < t < \infty), \tag{6}$$

where A is a given real constant square matrix. Of course the first thing to do is let

$$\vec{x} = P\vec{y} \tag{7}$$

and reduce system (6) to a canonical form

$$\frac{d\vec{y}}{dt} = B\vec{y} \;, \quad \text{where} \quad B = P^{-1}AP.$$

But here in general the well known Jordan canonical form will not be acceptable for we insist that P be a real matrix. The Russian text contained a mistake or misprints and my expectations were to correct this, which I did. The revised real Jordan canonical form, see [8], makes it possible to write out the general real solutions explicitly in terms of trigonometric functions, exponents and powers of t. From this the behavior can be readily determined in the large.

In retrospect I should have made it far clearer in paper [8] just how one would compute the appropriate transformation matrix P.

It would also have been well to point out in [8] that the explicit real solutions of the nonhomogeneous version

$$(8) \qquad \frac{d\vec{x}}{dt} = A\vec{x} + \vec{c}$$

of (6), where $\vec{c}$ is a given n-component constant vector, are easily worked out. To do this when A has an inverse, one sets

$$\vec{x} = \vec{y} - A^{-1}\vec{c}$$

and reduces the problem to the homogeneous case.

In the event that A does not have an inverse, transformation (7) would first be made and (8) would become

$$(9) \qquad \frac{d\vec{y}}{dt} = B\vec{y} + \vec{f} \;,$$

where B is in the suggested real revised Jordan canonical form and the $\vec{f} = P^{-1}\vec{c}$. Depending upon the number of blocks, say m, in B, system (9) would then break up into m separate systems of the type considered and each easily integrated and solved.

Wishing to generalize system (8), one might consider a real system

$$\frac{dx_i}{dt} = P_{im}(x_1, x_2, \ldots, x_n), \quad (i = 1, \ldots, n), \tag{10}$$

where each $P_{im}(x_1, x_2, \ldots, x_n)$ is a polynomial of degree m at most with real coefficients. Note that, if $m = 1$, (10) reduces to (8). However no systematic treatment of (10) is available. Even when we set $m = 2$, full details regarding the behavior of real solutions are not known. Indeed if $n \geq 3$ there seems no adequate vocabulary even for a qualitative description of the behavior of the solutions in the large. A number of special cases of (10) have, however, been considered, particularly in the Russian mathematical literature. We are faced here by a problem of extreme difficulty, but none the less important.

9. Rank reduction

In the equation

$$t^{r+1} \frac{dY}{dt} = \sum_{j=0}^{\infty} t^j A_j Y, \quad (r = 1, 2, \ldots), \tag{11}$$

where the indicated series converges for all sufficiently small $|t|$, A_j is a n by n constant square matrix and Y is the unknown n-component vector, the r by definition is the rank. H. Poincaré devised and J. Horn elaborated a method using quadratures for reducing the rank r to one, if at the outset r is larger than one; but, as G. D. Birkhoff pointed out, for applications their method was impractical. In my paper [38] a method for this reduction which entails a simple change of independent variable and an increase in the size of the matrices involved, brings about the desired reduction to rank one. Incidently in this paper [38] the first equation contains a misprint and it should have been written in form (11).

My expectation here was to utilize this result in the cut-off problem, presently to be described; but I have not been able to to this.

Inadvertently my rank reduction introduced extraneous solutions. D. A. Lutz in [39] has studied the interrelation of the solution of (11) and the extraneous solutions. He found, for example, that, if the original system (11) had a regular singularity at $t = 0$, so did the rank-reduced system. Lutz also supplied an example showing that the Poincaré method of rank reduction would not necessarily propagate the regular singular property from the original to the rank-reduced system. He also used the rank-reduced system to find new necessary conditions for the original system to have a regular singularity at $t = 0$.

10. The cut-off problem

The cut-off problem relates to a theorem of G. D. Birkhoff, [40], which states that, if one is given a differential equation

$$\frac{dX}{d\tau} = \tau^{q} \left(\sum_{k=0}^{\infty} \tau^{-k} A_k \right) X, \tag{12}$$

where $A_0 \neq 0$, q is an integer, the X and A_k's are square matrices with n rows and n columns and the indicated series is convergent for $|\tau| \geq \tau_0 > 0$, then there exists a transformation

$$X = \sum_{j=0}^{\infty} \tau^{-j} B_j Y,$$

where the square matrix B_0 is nonsingular and the series converges in some region $|\tau| \geq \tau_1 \geq \tau_0$, which will cut off the series in the equation (12) and convert (12) into the Canonical form

$$\frac{dY}{d\tau} = \tau^{q} \left(\sum_{j=0}^{s} \tau^{-j} C_j \right) Y .$$

All this is true; but Birkhoff thought that he had also proved that in no case is it necessary to take s greater than $q + 1$. However F. R. Gantmacher has produced a counter-example showing Birkhoff's bound $(q + 1)$ is wrong.

Since R. E. Langer was a Ph. D. student of G. D. Birkhoff and I was a Ph. D. student of Langer, I felt it my duty and great expectation that I could rescue Birkhoff's claim or at least correct his upper bound on s. All I was able to prove was that, if $q = -1$, the correct upper bound is not $(q + 1)$, but $(q + 2)$; see [11].

The rescue job was supposed to be done by proving that Birkhoff's bound on s would be correct if we would only admit some more general transformation, say one of the form

$$X = \tau^{p} \left(\sum_{j=0}^{\infty} \tau^{-j/q} B_j \right) Y,$$

for appropriate constants p and q with B_0 nonsingular. This result is only an expectation on my part. Nevertheless W. B. Jurkat and D. A. Lutz, [41], have already made some progress in this direction. Also see D. A. Lutz's lecture in these Proceedings.

11. Extensions and generalizations of the Lettenmeyer theorem

We now come to the research problem that I would liked to have solved and presented on this occasion rather than giving this expository lecture. First let me present the problem in a greatly simplified form. In introducing students to the subject of asymptotic series, one might begin with E. L. Ince's example, [42], which at first glance may seem to be very elementary; namely consider the equation

$$\frac{dw}{dz} = w + \frac{1}{z},$$

where temporarily we are interested in w and z as real variables and seek a solution approaching zero as $z \to \infty$ of the form

$$w_1(z) = \frac{c_1}{z} + \frac{c_2}{z^2} + \cdots + \frac{c_n}{z^n} + \cdots .$$

Formally one finds that

$$w_1(z) = \sum_{j=1}^{\infty} (-1)^{j-1}(j-1)! \ / z^j ;$$

but unfortunately this series diverges for all finite values of z. However the solution we want does take the form

$$w_2(z) = - \int_z^{\infty} \frac{e^{z-\sigma} d\sigma}{\sigma} , \quad z > 0,$$

and, after integrating n times by parts, one finds

$$w_2(z) = \sum_{j=1}^{n} (-1)^{j-1}(j-1)! \ z^{-j} + R_n(z),$$

where the remainder term

$$R_n(z) = (-1)^n \ n! \int_z^{\infty} \frac{e^{z-\sigma} d\sigma}{\sigma^{n+1}} .$$

An easy estimate shows that

$$| R_n(z) | < \frac{n!}{|z|^{n+1}} , \quad \text{if} \quad z > 0.$$

Thus $w_1(z)$ is an asymptotic expansion representing our solution. If we permit w and z to become complex variables and extend our solution analytically into the complex z-plane, an estimate by K. O. Friedrichs, [43], shows that, if $|z| > 0$ and $|\arg z| \leq \pi$, then

$$|R_n(z)| \leq n! \ (n\pi + \pi + 2) / |z|^{n+1} .$$

Indeed the asymptotic representation $w_1(z)$ is valid in the sector $|\arg z| < \frac{3\pi}{2}$, which is peculiar unless one realizes that the solution $w_2(z)$ is a multiple valued function of the form

$$w_2(z) = e^z \left[\gamma + \log z - z + \frac{z^2}{2(2!)} - \cdots + \frac{(-1)^n z^n}{n(n!)} + \cdots \right] ,$$

where γ is Euler's constant and the indicated series converges for all finite values of z.

If one wanted a very accurate value of $w_2(z)$ for a fixed very large value of z, one could not use the divergent asymptotic series solution. Nevertheless the divergent series can be summed. In fact by a change of variable

$$w_2(z) = -\int_z^{\infty} \frac{e^{z-\sigma}\,d\sigma}{\sigma} = -\int_0^1 \frac{t^{z-1}\,dt}{1-\log t}$$

and this last Laplace integral representation of our solution, according to L.M. Milne-Thompson, [44], can be expanded into a factorial series

(13) $$w_2(z) = -\sum_{n=0}^{\infty} \frac{n!\,a_n}{z(z+1)\cdots(z+n)} ,$$

where

$$a_0 = 1 \text{ and } a_n = -\sum_{j=0}^{n} a_j/(n+1-j),$$

and the series in (13) is convergent for $\operatorname{Re} z > 0$.

One might be tempted to ask the students to get an analogous convergent factorial series representation for the portion of the z-plane where $\operatorname{Re} z \leq 0$ for large values of $|z|$. But not being sure how to do this, instead, one might ask them in their homework to parallel my presentation using the equation

$$\frac{dw}{dz} = w + \frac{1}{z} + \frac{1}{z^2} .$$

However this time the formal series solution running in powers of $1/z$ is convergent. So this brings us to a new problem.

Given

$$\frac{dw}{dz} = w + a_0 + \frac{a_1}{z} + \cdots + \frac{a_n}{z^n} + \cdots ,$$

where the series on the right converges is $|z|$ is large enough, say $|z| \geq z_0 > 0$, when does the formal asymptotic series solution

$$w_3 = b_0 + \frac{b_1}{z} + \ldots + \frac{b_n}{z^n} + \ldots$$

converge ? My last class came to the conclusion that the series w_3 converges if and only if the convergent series

$$\sum_{m=1}^{\infty} \frac{(-1)^m a_m}{(m-1)!} = 0.$$

This brings us to the end of my talk and a problem for the future; namely given the equation

$$\text{(14)} \qquad \frac{dW}{dz} = \sum_{j=0}^{\infty} z^{-j} A_j W + \sum_{j=1}^{\infty} z^{-j} B_j$$

with a formal solution

$$\text{(15)} \qquad W(s) = \sum_{j=0}^{\infty} z^{-j} C_j ,$$

where the $W(s)$, B_j , and C_j are n-component vectors, and the A_j are n by n given constant matrices and both series in (14) converge if $|z| \geq z_0 > 0$, what are the necessary and sufficient conditions that (15) converge ?

Perhaps W. A. Harris, Jr., Y. Sibuya, and L. Weinberg can help us. See reference [45].

References

1. H. L. Turrittin. "Asymptotic solutions of certain ordinary differential equations associated with multiple roots of the characteristic equations," Amer. Jour. of Math., 58 (1936) 364-376.

2. ______________ "Asymptotic distribution of zeros for certain exponential sums," Amer. Jour. of Math., 66 (1944) 199-228.

3. H. L. Turrittin. "Stokes multipliers for asymptotic solutions of a certain differential equation," Trans, Amer. Math. Soc., 68 (1950) 304-329.

4. ______________ "Asymptotic expansions of solutions of systems of ordinary linear differential equations containing a parameter," Annals of Math. Studies, vol. II, No. 29, Princeton: Princeton Univ . Press, (1952) 81-116.

5. ______________ "Convergent solutions of ordinary linear homogeneous differential equations in the neighborhood of an irregular singular point," Acta Mathematica, 93 (1955) 27-66.

6. ______________ " A peculiar solution of a modified Duffing's equation" (with W. J. A. Culmer), Ann Mat. Pura Appl. Ser. IV, 44 (1957) 23-33.

7. ______________ "Standardization and simplification of systems of linear differential equations involving a turning point" (with W.A. Harris, Jr.), SIAM J. Appl. Math 7 (1959) 316-324.

8. ______________ "Linear differential or difference equations with constant coefficients," Amer. Math. Monthly, 66 (1959) 869-875.

9. ______________ "The formal theory of systems of irregular homogeneous linear difference and differential equations," Boletin de la Sociedad Math. Mexicana, (1960) 255-264.

10. ______________ " A canonical form for a system of linear difference equations," Ann. Mat. Pura Appl. Ser. IV, 58 (1962) 335-358.

11. ______________ "Reduction of ordinary differential equations to the Birkhoff canonical form," Trans. Amer. Math. Soc., 107 (1963) 485-507.

12. ______________ " Solvable related equations pertaining to turning point problems," Asymptotic solutions of differential equations and their applications, edited by Calvin H. Wilcox, John Wiley &Sons, (1964) 27-52.

13. H. L. Turrittin. "Stokes multipliers for the differential equation $\frac{d^n y}{dx^n} - \frac{y}{x} = 0$," Funkcialaj Ekvacioj, 6 (1964) 37-46.

14. ____________ "Convergent solutions of ordinary linear nonhomogeneous differential equations," Funkcialaj Ekvacioj, 12 (1969) 7-21.

15. R. E. Langer. "The solution of the differential equation

$$v''' + \lambda z v' + 3\mu\lambda^2 v = 0,"$$

Duke Math. J. 22 (1955) 525-542.

16. ____________ "On the asymptotic forms of ordinary linear differential equations of the third order in a region containing a turning point," Trans. Amer. Math. Math. Soc., 80 (1955) 93-123.

17. ____________ "The solutions of a class of ordinary linear differential equations of the third order in a region containing a multiple turning point," Duke Math. J. 23 (1956) 93-110.

18. ____________ "On the asymptotic solutions of a class of ordinary differential equations of the fourth order with a special reference to an equation of hydrodynamics," Trans. Amer. Math. Soc. 84 (1957) 144-191.

19. Y. Sibuya. "Simplification of a linear ordinary differential equation of the n-th order at a turning point," Arch. Rat. Mech. Anal., 13 (1963) 206-221.

20. ____________ "On the problem of turning points for a system of linear ordinary differential equations of higher orders," Proc. of Symposium Math. Res. Ctr., U.S. Army, Univ. Wisconsin, Madison, Wisc. (1964) 145-162, Wiley, N.Y. 1964.

21. W. Wasow. Asymptotic expansions for ordinary differential equations, Interscience Publishers, 1965, Chap. 8.

22. ________ "On turning point problems for systems with almost diagonal coefficient matrix," Funkcialaj Ekvacioj, 8 (1966) 143-170.

23. ________ "The central connection problem at turning points of linear differential equations," Analytic theory of differential equation, Lecture

notes in Mathematics, #183, Springer-Verlag, 1971, (Proc. of a conference at Western Mich. Univ.), 158-164.

24. L.A. MacColl. "On the distribution of the zeros of sums of exponentials of polynomials," Trans. Amer. Math. Soc., 36 (1934) 341-360.

25. R.E. Langer. "The asymptotic location of the roots of a certain transcendental equation," Trans. Amer. Math. Soc., 31 (1929) 837-844.

26. J.G. Van der Corput. "Rhythmische Systeme," Acta Math. 59 (1932) 209-328.

27. J.B. McLeod. "On the distribution of eigenvalues of an n-th order equation," Quart, J. Math. Oxford (2), 17 (1966) 112-131.

28. B.L.J. Braaksma. "Asymptotic analysis of a differential equation of Turrittin," SIAM J. Math. Anal, 2 (1971) 1-16.

29. H.L. Turrittin. "Stokes multipliers for the equation $\frac{d^3y}{dx^3} - \frac{y}{x^2} = 0$," Lecture notes in mathematics, #183, Analytic theory of differential equations, (Proc. of Conf. at Western Mich. Univ. (1970) 145-157.

30. H. Scheffé. "Linear differential equations with two term recurrence formulas," J. Math.& Phys., 21 (1942) 240-249.

31. W. Loud & K.W. Blair. "Periodic solutions of $x'' + cx' + g(x) = Ef(t)$ under variation of certain parameters," SIAM J. Appl. Math. 8 (1960) 74-101.

32. W. Loud. "Periodic solutions of second order differential equations of Duffing type," Proc. U.S.-Japan Seminar, (1967) 199-224.

33. ________ "Branching phenomena for periodic solutions of nonautonomous piecewise linear systems," International Jour. of Nonlinear Mechanics, 5 (1969 352-368.

34. W.J.A. Culmer & W.A. Harris, Jr. "Convergent solutions of ordinary linear homogeneous difference equations," Pac. Jour. Math., 13 (1963) 1111-1138.

35. W.A. Harris, Jr. &S. Tanaka. "On difference equations containing a parameter," Publ. of Research Inst. for Math. Sciences, Kyoto Univ., Ser. A, 2 (1966), 5-16.

36. R. Gérard & A.H.M. Levelt. "Invariants mesurant l'irregularité en un point singulier des systemes d'équations différentielles linéariares," Notes published by the Institut de Recherche Mathématique Avancée de Strasbourg, 1972.

37. A.H.M. Levelt. "Formal theory of irregular singular points," (to appear), 1972.

38. H.L. Turrittin." Reducing the rank of ordinary differential equations," Duke Math. Jour., 30 (1963) 271-274.

39. D.A. Lutz. "On the reduction of rank of linear differential systems," U.S. Army Math. Research Center, Univ. of Wisc., Madison, Wisc., Report #1097, Aug., 1970, 1-18.

40. G.D. Birkhoff, "Equivalent singular points of ordinary linear differential equations," Math. Ann. 74 (1913), p. 136.

41. W.B. Jurkat & D.A. Lutz. "Birkhoff reduction of two-dimensional linear differential systems at a singular point," U.S. Army Math. Research Center, Univ. of Wisc., Madison, Wisc., Report #1062, (1970), 1-33.

42. E.L. Ince. Ordinary Differential Equations, 1927 edition, p. 174.

43. K.O. Friedrichs. Special topics in analysis, New York Univ. Notes, 1953, page B-9.

44. L.M. Milne-Thompson. The calculus of finite differences, 1951, pp. 289-290.

45. W.A. Harris, Jr., Y. Sibuya &L Weinberg. "Holomorphic solutions of Linear differential systems at singular points," Arch. Rat. Mech. &Anal, 35 (1969), 245-248.

46. F. Lettenmeyer. " Über die an einer Unbestimmtheitsstelle regulären Lösungen eines Systemes homogener linearen Differentialgleichunghen," S. - B. Bayer. Akad. Wiss. Munchen Math. - nat. Abt. (1926) 287-307.

Admissibility and the Integral Equations of Asymptotic Theory

H. E. Gollwitzer

1. Introduction

The determination of the asympototic properties of solutions of ordinary differential equations generally involves procedures which lead one to consider integral operators and certain types of integral equations. One way to study the various stability properties of solutions of the nonlinear vector differential equation

$$(1.1) \qquad y' = A(x)y + f(x, y), \quad x \geq 0$$

is to examine the nonsingular integral equation

$$(1.2) \qquad y(x) = y_0(x) + \int_0^x K(x,t)f(t, y(t))dt$$

where y_0 and K involve the fundamental solution matrix of the unperturbed equation. This well known procedure, which dates back of some investigations of Liouville, has been refined over the years to the point where it is recognized as an important tool in stability theory. Along these lines we only mention the book by Coppel [1].

The study of (1.1) in the analytic case also leads to equations of the type indicated in (1.2), However, the nature of the questions one asks about solutions of (1.2) differ greatly from those indicated in the book by Coppel. In fact, one is generally led to consider singular integral equations wherein the kernel K and the free term y_0 have singularities at the endpoint $x = 0$. These singularities occur naturally as part of a program to establish rather precise asymptotic estimates for solutions of differential equations in a region or sector near a singular point. We might say a word or two about this program. The data A, f

for equation (1.1), and possibly other auxiliary conditions, usually suggest potentially useful asymptotic forms or approximations for the solutions of (1.2), but this step is quite delicate and one must proceed with caution lest the true asymptotic behavior go unnoticed. Some idea of the difficulties inherent in this step can be found in a talk given by Professor Turrittin a few years ago [11]. Although a great deal ofthought goes into the selection of a function or asymptotic series which might describe some or all of the solutions of(1.1) there remains the problem of showing that the approximation is appropriately close to the desired solution. It is then of some importance to have a catalogue of results on integral equations which point out the types of theorems one might expect concerning existence and/or uniqueness. Such a program is in harmony with a procedure outlined by Langer [6] some years age.

In the last few years, Massera and Schaffer [8], Corduneanu [2] and others isolated an important concept called admissibility which is associated with the integral operators one encounters in stability problems. Some extensions were recently given in [5]. Our purpose in this paper is to discuss a few theorems concerning the global behavior of solutions of integral equations. In so doing, we will indicate how the work of Erdélyi [4] can be phrased in terms of admissibility. Also, we will briefly mention some older results which deserve more attention.

2. Admissibility

This section is devoted to a brief discussion of a few known results associated with the concept of admissibility. Let X and Y be given subspaces of a fixed Frechet space F which admit norms $|\cdot|_1$ and $|\cdot|_2$, respectively. Assume that X and Y are Banach spaces under their respective norms, and X and Y are sequentially stronger than F in then sense that if a sequence in X, or a sequence in Y converges in norm, then the sequence also converges in the topology of F.

Definition 1. Let T be a continuous linear operator on F. Then T is said to be admissible with respect to the pair (X, Y) if and only if $TX \subset Y$.

Let X and Y be as above and assume that T is a continuous linear operator on F.

Theorem 1. *If (X, Y) is admissible with respect to T, then T is a continuous linear operator from X into Y and hence $|T| \equiv \sup \{ |Tx|_2 : |x|_1 = 1 \}$ is finite.*

Further details and proofs are given by Miller in [10].

If these results are to be useful, one must have at his disposal an adequate description of a large number of the function spaces which occur in practice. Many of these function spaces are motivated by the applications, and a few of them are described in the following paragraphs. More general possibilities are described in [5], where further details and proofs of the results we will discuss in this section can be found.

For the remainder of this section we will use the notation $I = I(\omega) = \{t: 0 \leq t < \omega \leq \infty\}$ and let E^n denote the inner product space of real n-tuples $u = (u^1, \ldots, u^n)$, with the standard inner product $(u, v) = \Sigma u^i v^i$ and norm $\|u\| = (u, u)^{1/2}$. We will also consider the related inner product space E^{nxl} of real nxl matrices with the standard inner product. The complex versions of E^n and E^{nxl} can be easily incorporated into our work, but this will be left to the reader. We first define a weighted function space C_G of continuous functions. For each t in I, let G(t) denote a linear transformation on E^n with the property that $|G(t)|$ is uniformly bounded on compact subsets of I, where $|G(t)| = \sup_{\|u\|=1} \|G(t)u\|$. Let N(t) and R(t) denote the null space and the range of G(t), and let $M^\perp$ be the orthogonal complement of any subspace M of E^n. If P(t) is the orthogonal projection of E^n onto R(t), define $G_{-1}(t)$: $R(t) \to N^{\perp}(t)$ to be the inverse of the restriction of $G(t)$ to $N^{\perp}(t)$. We define $P(t) = 0$ if $R^{\perp}(t) = E^n$. Also, extend $G_{-1}(t)$ linearly by sending $R^{\perp}(t)$ into the zero vector and call the resulting transformation $G_{-1}(t)$. A natural function space of continuous functions from I into E^n is the one associated with the topology of uniform convergence on compact subsets of I, and we call it C_c or $C_c(I, E^n)$.

Definition 2. A function u in C_c is said to be in $C_G(\equiv C_G(I, E^n))$ if $P(t)\, u(t) = u(t)$ for each t in I and

$$|u|_G \equiv \sup_I \| G_{-1}(t) u(t) \|$$

is finite.

If u is in C_c and $P(t)\, u(t) = u(t)$ for each t in I, then $u(t) = G(t)G_{-1}(t)u(t)$. This definition offers a natural generalization of the Landau order symbol if we write

$$\text{"}u(x) = O(G(x)) \text{ as } x \to \omega\text{"}$$

to mean that u is in $C_G(J, E^n)$ for some interval $J = [a, \omega)$. The following theorem is proved in [5].

Theorem 2. The space C_G is a Banach subspace of C_c which is sequentially stronger than C_c.

In some situations it is more natural to work with measurable rather than continuous functions. This leads to the definition of L_G^{∞}. For each t in I let $G(t)$ be a linear transformation on E^n and $e_1, \ldots, e_n$ the standard ordered orthonormal basis in E^n. By definition the transformation is Lebesgue measurable on I if the real valued functions $(G(t)e_j, e_i)$, $i, j = 1, \ldots, n$, are Lebesgue measurable on I. This is of course equivalent to asking that the matrix of $G(t)$ be such that it entries are measurable. It was shown in [5] that if $G_{-1}(t)$ is defined according to the plan described in the paragraph which preceded Definition 2 and $G(t)$ is measurable on I, then $G_{-1}(t)$ is measurable on I. This fact assures us that $\|G_{-1}(t)u(t)\|$ is a measurable function on I.

<u>Definition 3</u>. Let $G(t)$ be a linear transformation on E^n which is measurable on I. A measurable function u from I into E^n is said to be in L_G^∞ $(\equiv L_G^\infty(I, E^n))$ if $P(t)u(t) = u(t)$ almost everywhere on I and

$$|u|_{G^\infty} \equiv \operatorname*{ess\ sup}_I \|G_{-1}(t)u(t)\|$$

is finite.

It follows immediately that L_G^∞ is a Banach space with norm $|\cdot|_{G^\infty}$.

We can also characterize the admissibility of a pair (C_g, C_G) with respect to a linear integral operator T generated by a continuous function $K(t, s)$ which, for each (t, s), is a linear transformation on E^n. In addition to the usual assumptions concerning g and G, we assume that g is continuous and $P(t)$ is the projection associated with $G(t)$ in the definition of C_G.

<u>Theorem 3.</u> <u>The pair</u> (C_g, C_G) <u>is admissible with respect to the linear operator</u> T <u>given by</u>

$$Tu(t) = \int_0^t K(t, s)u(s)ds, \quad t \in I$$

<u>if and only if</u>

(i) $$P(t)K(t, s)g(s) = K(t, s)g(s) \quad \text{for } 0 \leq s \leq t$$

(ii) $$\int_0^t |G_{-1}(t)K(t, s)g(s)|\, ds \leq A$$

<u>where</u> A <u>is some constant which depends only on</u> g, G <u>and</u> K.

Let $K(t, s)$ denote an $n \times n$ matrix function with continuous entries $K_{ij}(t, s)$ for $0 \leq s \leq t$. If f is a continuous matrix function of size $n \times 1$ with real entries, then the Volterra integral equation

$$\text{(L)} \qquad u(t) = f(t) + \int_0^t K(t, s)u(s)ds$$

has a unique solution given by

$$u(t) = f(t) + \int_0^t k(t, s)f(s)ds$$

where the resolvent kernel $k(t, s)$ is continuous and of size $n \times n$.

By definition the pair (C_g, C_G) is admissible for (L) if u is in C_G whenever f is in C_g. The matrices g and G are as described in Theorem 3.

Theorem 4. The pair (C_g, C_G) is admissible for (L) if and only if

(i) $P(t)g(t) = g(t)$ on I;

(ii) $P(t)k(t, s)g(s) = k(t, s)g(s)$ for $0 \leq s \leq t$

and

$$\text{(iii)} \qquad |G_{-1}(t)g(t)| + \int_0^t |G_{-1}(t)k(t, s)g(s)|\, ds \leq A$$

where A is a constant which depends only on g, G and K.

3. An Associated Equation

The formulation of the theorem in Section 2 was specifically designed to suggest that a suitable normalization of the problem leads one to consider an associated equation whose solutions are uniformly bounded on I. In order to illustrate this point, we take as an example the simple nonsingular integral equation

$$(3.1) \qquad u(x) = f(x) + \int_0^x K(x, t)u(t)dt$$

where the nxn matrix function $K(x,t)$ is continuous for $0 \le t \le x < \infty$ and f, u are nx1 matrix functions which are continuous on $I = [0, \infty)$. The unique continuous solution is given by

$$u(x) = f(x) + \int_0^x k(x,t)f(t)dt$$

where k is the resolvent kernel associated with K. Let $G(t)$ be an nxn matrix function which is continuous on I and define $G_{-1}(t)$, $P(t)$ as in Section 2. We note that $G_{-1}(t)$ is measurable on I, and (C_G, C_G) is admissible for (3.1) if and only if $P(x)k(x,t)G(t) = k(x,t)G(t)$ when $0 \le t \le x$ and

$$\int_0^x |G_{-1}(x)k(x,t)G(t)|\, dt \le A .$$

Although these conditions characterize the admissibility of (C_G, C_G) for (3.1), it is usually more desirable to have estimates which directly involve K and not k. If some solution u is in C_G, then the function v defined by $v(t) = G_{-1}(t)u(t)$ is measurable on I and $u(t) = G(t)G_{-1}(t)u(t) = G(t)v(t)$. Upon multiplying the members of (3.1) by $G_{-1}(t)$, we see that v is a solution of

$$v(x) = G_{-1}(x)f(x) + \int_0^x G_{-1}(x)K(x,t)G(t)v(t)dt \tag{3.2}$$

which is uniformly bounded on I. Since sufficient conditions for integral equations to have bounded solutions have been known for years, the integral equation (3.2) suggests a way of locating conditions which imply that u is in C_G whenever f is in C_G.

Theorem 5. *Suppose that*

(i) $|G_{-1}(x)|$ *is uniformly bounded on compact subsets of* $I = [0, \infty)$;

(ii) $P(x)K(x,t)G(t) = K(x,t)G(t)$ *for* $0 \le t \le x$;

(iii) $\int_0^x |G_{-1}(x)K(x,t)G(t)|\,dt \le A, \; x \ge 0;$

(iv) <u>for some</u> $x_0 \ge 0$

$$\int_{x_0}^{x} |G_{-1}(x)K(x,t)G(t)|\,dt \le \alpha < 1, \; x \ge x_0.$$

<u>Then</u> (C_G, C_G) <u>is admissible for</u> (3.1).

<u>Proof</u>. We must first show that a solution u satisfies $P(x)u(x) = u(x)$ on I whenever the continuous function f satisfies the same algebraic condition with u replaced by f. Suppose that it has been shown that $P(x)k(x,t)G(t) = k(x,t)G(t)$ holds when x and t satisfy $0 \le t \le x$. The resolvent formula

$$u(x) = f(x) + \int_0^x k(x,t)f(t)dt$$

together with the known relationships

$$P(x)f(x) = f(x) = G(x)G_{-1}(x)f(x)$$

and

$$P(x)k(x,t)G(t) = k(x,t)G(t)$$

clearly imply that $P(x)u(x) = u(x)$ holds for all x in I. It remains to establish the identity displayed in the last formula line. The resolvent k is given by the series $k(x,t) = \sum_1^{\infty} K_i(x,t)$, where $K_1(x,t) = K(x,t)$ and

$$K_i(x,t) = \int_t^x K(x,s)K_{i-1}(s,t)ds, \quad i \ge 2.$$

The hypotheses on K imply that

$$K(x,t)G(t) = G(x)\,G_{-1}(x)K(x,t)G(t)$$

and hence

$$P(x)K_2(x,t)G(t) = P(x) \int_t^x K(x,s)G(s)G_{-1}(s)K(s,t)G(t)ds$$

$$= \int_t^x K(x,s)G(s)G_{-1}(s)K(s,t)G(t)ds = K_2(x,t)G(t)$$

An induction argument shows that

$$P(x)K_i(x,t)G(t) = K_i(x,t)G(t), \quad i \geq 2$$

and hence the previous identity holds with K_i replaced by k.

We now proceed to show that u is in C_G whenever f is in C_G. The identity $P(x)u(x) = u(x)$ permits us to write

$$v(x) = G_{-1}(x)f(x) + \int_0^x G_{-1}(x)K(x,t)G(t)v(t)dt$$

where $v(x) = G_{-1}(x)u(x)$ is measurable and uniformly bounded on compact subsets of I. If M denotes the bound for v on $[0, x_0]$, then the previous formula line leads to the estimate

$$(3.3) \quad |v(x)| \leq |f|_G + M_0 \int_0^{x_0} |G_{-1}(x)K(x,t)G(t)|dt + \int_{x_0}^x |G_{-1}(x)K(x,t)G(t)|\,|v(t)|dt$$

when $x \geq x_0$. The hypotheses of the theorem imply that the first two terms in the right member of this inequality are bounded by some constant N on I. Let w(x) denote the finite least upper bound of $|v(s)|$ on $[x_0, x]$. Fix $x > x_0$ and let s vary between x_0 and x. We conclude from (3.3) that

$$|v(s)| \leq N + w(x) \int_{x_0}^s |G_{-1}(s)K(s,t)G(t)|dt \leq N + \alpha\, w(x)$$

and hence $w(x) \leq N(1 - \alpha)^{-1}$. Since x was quite arbitrary and v was bounded on $[0, x_0]$, we see that $|u|_G = \sup_I |v(x)|$ is finite and the theorem is complete.

Hypothesis (i) was used to show that $v(x) = G_{-1}(x)u(x)$ was bounded on compact subsets of I, and it is interesting that a slight modification in hypothesis (iii) permits one to omit (i).

Theorem 6. Suppose that

(i) $P(x)K(x,t)G(t) = K(x,t)G(t)$ for $0 \le t \le x$;

(ii) $\int_0^x |G_{-1}(x)K(x,t)G(t)|\, dt \le A < 1,\ x \ge 0.$

Then (C_G, C_G) is admissible for (3.1).

Proof. The hypotheses of the theorem and the contraction mapping theorem imply that the integral equation

$$v(x) = h(x) + \int_0^x G_{-1}(x)K(x,t)G(t)v(t)dt$$

has a bounded measurable solution on I. Let v be the bounded solution of

$$v(x) = G_{-1}(x)f(x) + \int_0^x G_{-1}(x)K(x,t)G(t)v(t)dt$$

when f is in C_G and define $u(x) = G(x)v(x)$. Upon multiplying the previous formula line by $G(x)$, we see that u satisfies the original integral equation(3.1). Since u is bounded on compact subsets of I, we conclude that u must be the unique continuous solution of (3.1) and

$$|u|_G = \sup_I \|G_{-1}(t)u(t)\| \le \sup_I \|v(t)\|$$

is finite. The proof is complete

4. A Perturbation Problem

The following perturbation problem has been discussed by several writers: If the solution u of

(4.1) $$u(x) = f(x) + \int_0^x K(x,t)u(t)dt, \qquad x \in I = [0, \infty)$$

is bounded whenever f is bounded on I, what restrictions should one impose on $K_0(x,t)$ in order to ensure the same conclusion for the solution of

(4.2) $$u(x) = f(x) + \int_0^x [K(x,t) + K_0(x,t)]\, u(t)\, dt$$

whenever f is bounded on I ? This problem has been discussed by Corduneanu [3] while Miller [9] used a different approach to study perturbation problems. A detailed discussion was also given by Miller in [10]. The arguments presented here are probably not new, but the author has been unable to locate a similar discussion in the literature. For the sake of convenience, we assume that the nxn matrix functions K and K_0 are continuous and the nxl matrix function f is continuous on I. If k(x, t) is the continuous resolvent assoicated with K, then when u is a solution of (4.2) and v is defined as

$$v(x) = u(x) - \int_0^x K(x,t)u(t)dt$$

we see that

(4.3) $$u(x) = v(x) - \int_0^x k(x,t)v(t)dt$$

and v is a solution of

(4.4) $$v(x) = f(x) + \int_0^x [K_0(x,t) - \int_t^x K_0(x,s)k(s,t)ds]\; v(t)\; dt$$

The hypotheses concerning (4.1) together with Theorem 4 imply that $\int_0^x |k(x,t)|dt$ is bounded on I, and hence one concludes from (4.3) that u will be bounded if v is bounded on I. Although the kernel in (4.4) is somewhat complicated, it contains the basic information one needs in order to be able to describe sufficient conditions for v, and hence u, to inherit the boundedness property of f. The most general conditions deal with the resolvent kernel k_0 associated with the kernel

$$K_0(x,t) - \int_t^x K_0(x,s)k(s,t)ds, \quad 0 \le t \le x.$$

Since a solution v of (4.4) is given by the resolvent formula

$$v(x) = f(x) - \int_0^x k_0(x,t)f(t)dt,$$

the next theorem needs no formal proof.

Theorem 7. *If* $\int_0^x |k_0(x,t)|dt$ *is bounded on* I, *then the solution* u *of* (4.2) *is bounded whenever* f *is bounded on* I.

With Theorem 5 in mind, we state the following corollaries.

Corollary 1. *If*

$$\int_0^x \left| K_0(x,t) - \int_t^x K_0(x,s)k(s,t)ds \right| dt$$

is bounded on I *and if, for some* $x_0 \ge 0$, *it can be established that*

$$\int_{x_0}^x \left| K_0(x,t) - \int_t^x K_0(x,s)k(s,t)ds \right| dt \le \alpha < 1, \quad x \ge x_0$$

then the solution u *of* (4.2) *is bounded whenever* f *is bounded on* I.

These conditions are sufficient to ensure that $\int_0^x |k_0(x,t)| dt$ is bounded on I, and hence the result follows from the theorem.

Corollary 2. *If* $\int_0^x |K_0(x,t)| dt \le \beta$ *and* β *is sufficiently small, then the solution* u *of* (4.2) *is bounded whenever* f *is bounded on* I.

This result follows from Corollary 1 after we observe that a smallness condition on β eventually forces the hypotheses in that corollary. The last corollary was proved by Corduneanu in [3].

5. Results of Erdélyi and Love

In a series of papers, Erdélyi [4] gave a detailed discussion of a singular nonlinear integral equation of the form

$$u(x) = f(x) + \int_c^x K(x, t, u(t))\, dt$$

which served as a good model for the integral equations one generally encounters in the asymptotic theory of ordinary differential equations. The point c is the endpoint of a possibly unbounded interval I and the solutions are potentially discontinuous at $x = c$ or on a subset of I. Although the functions considered by Erdélyi had their values in some Banach space, we will only consider the simpler case when u, f and K have their values in E^{nxl}. There is no real loss of generality here in considering instead the singular integral equation

$$u(x) = f(x) + \int_x^\infty K(x,t, u(t))dt, \quad x \geq a \tag{5.1}$$

where f is an nxl matrix function which is measurable on $I = [a, \infty)$ and K satisfies the standard hypotheses

$K(x, t, u)$ is defined for all u in E^{nxl} and for values of x and t which satisfy $a \leq t \leq x$;

$K(x, t, u)$ is plane measurable in (x, t) for each u in E^{nxl};

$K(x, t, u)$ is continuous in u for almost all (x, t) in the region determined by the condition $a \leq t \leq x$.

We are now in a position to discuss a typical case of a theorem proved by Erdélyi [4], but first some additional assumptions concerning (5.1) must be described. The assumptions are as follows:

(i) there exists a measurable and almost everywhere finite function and an integrable function g on I such that

$$k(x,t)\,\eta(t) \leq \eta(x)\,g(t)$$

holds almost everywhere for $a \leq t \leq x$ and k is a measurable function which is related to K by means of the condition

$$\|K(x,t,u) - K(x,t,v)\| \leq k(x,t)\,\|u-v\|\,;$$

(ii) f is measurable on I and $\|f(x)\| \leq \eta(x)$ holds almost everywhere on I;

(iii) $K(x,t,0) = 0$ when x and t satisfy the condition $a \leq t \leq x$.

Under these assumptions Erdélyi has shown that a solution of (5.1) can be constructed by successive approximations, with $u_{-1}(x) = 0$, $u_0(x) = f(x)$,

$$u_n(x) = f(x) + \int_x^\infty K(x,t,u_{n-1}(t))\,dt, \quad n = 1,2,\ldots$$

The solution u is given by the series

$$u(x) = \Sigma[u_k(x) - u_{k-1}(x)]$$

which converges absolutely on I, and as part of the existence theorem one obtains the useful estimates

$$\|u(x) - u_n(x)\| \leq \eta(x)\; e_{n+1}(g_1(x))$$

where

$$e_n(z) = \exp(z) - \sum_0^{n-1} z^k/k\,! \quad ; \quad g_1(x) = \int_x^\infty g(t)dt.$$

The details of these computations can be found in [4], where it is also shown that the solution u is the only solution which satisfies the condition $\|u(x)\| \leq \eta(x)$ almost everywhere on I.

The concept of admissibility is present in this development, and it is perhaps worthwhile to isolate some important observations.
The condition $\|f(x)\| \leq \eta(x)$ suggests that we consider the weighted Banach space L_G^∞, where $G(t) = \eta(t)E$ and E is the nxn identity matrix. The projection $P(t)$ associated with L_G^∞ is given by $P(t) = \chi(t)(1 - \chi(t))^{-1} E$, where χ is the characteristic function of the set $\{t : \eta(t) > 0\}$. Also, $G_{-1}(t) = \chi(t)(1 - \chi(t) - \eta(t))^{-1} E$ is certainly measurable on I and the pair (L_G^∞, L_G^∞) is admissible with respect to the integral operator T given in (5.1). If u is in L_G^∞, then $u(t) = G(t)v(t)$ and v is essentially bounded on I. Thus,

$$\|Tu(x)\| \leq \int_x^\infty \|K(x,t,u(t))\|dt \leq \int_x^\infty k(x,t)\|u(t)\|\,dt \leq M\,\eta(x)\int_x^\infty g(t)\,dt$$

where M is the essential supremum of v. We conclude that $P(x)Tu(x) = Tu(x)$ on I and the previous inequality can be written as

$$\|G_{-1}(x)Tu(x)\| \leq |u|_{G^\infty} \int_x^\infty g(t)dt \leq |u|_{G^\infty} \int_a^\infty g(t)dt ,$$

almost everywhere on I. Hence it is true that $Tu(x)$ is in L_G^∞ whenever u is in L_G^∞. An important part of Erdélyi's work dealt with the quantitative condition

$$k(x,t)\,\eta(t) \leq \eta(x)g(t)$$

which enable him to derive several important and useful estimates. However, part of Erdélyi's work was anticipated by Love [7] who, in 1914, published an interesting paper on singular integral equations and presented many interesting ideas which in some sense even anticipated the integral equation version of admissibility. It is perhaps unfortunate that this paper has gone unnoticed for so many years.

The viewpoint afforded us through the concept of admissibility puts us in a position to formulate a general theorem which includes the results of Erdélyi just mentioned. The following theorem can be easily modified to include the results given in [4]. We are interested in the integral equation (5.1) and an $n \times n$ matrix function $G(t)$ which is measurable on $I = [a, \infty)$. This permits us to define L_G^∞ and the associated projection $P(t)$.

Theorem 8. Suppose that

(i) $K(x,t,0) = 0$ and $P(x)K(x,t,G(t)u) = K(x,t,G(t)u)$ for almost all x and t and all u in $E^{n \times 1}$;

(ii) $\| G_{-1}(x) [K(x,t,G(t)u) - K(x,t,G(t)v)] \| \leq g(t) \| u-v \|$

almost everywhere in (x,t) when $a \leq t \leq x$ and the vectors u and v are arbitrary in $E^{n \times 1}$;

(iii) g is integrable on $[a, \infty)$;

Then (5.1) has a unique solution u in L_G^∞ which also satisfies the estimate

$$\| G_{-1}(x) [u(x) - f(x)] \| \leq |u|_{G\infty} \int_x^\infty g(t)\, dt$$

almost everywhere on I.

This theorem shows that an appropriate weighted difference of u and f approaches zero as x approaches infinity. The proof of the theorem is tedious but not difficult. If v is an essentially bounded solution of

$$(5.2) \qquad v(x) = G_{-1}(x)f(x) + \int_x^\infty G_{-1}(x)K(x,t,G(t)v(t))dt$$

on I, then it is not difficult to check that $u(x) = G(x)v(x)$ is a solution of (5.1). By construction, $P(x)u(x) = u(x)$ and $|u|_{G\infty} < \infty$. If w is another solution in L_G, define $w_0(x) = G_{-1}(x)w(x)$ and takes differences in (5.1), (5.2) to obtain

$$\|v(x) - w_0(x)\| \leq \int_x^\infty g(t) \, \| v(t) - w_0(t)\| \, dt$$

almost everywhere for $x \geq a$. A singular version of Gronwall's inequality leads to the conclusion that $v(x) = w_0(x)$ almost everywhere on I. Thus, u is unique in the sense that $|u - w|_{G^\infty} = 0$. The existence of a solution v of (5.2) is established by the method of successive approximations. If u is the solution of (5.1), then

$$G_{-1}(x) [u(x) - f(x)] = \int_x^\infty G_{-1}(x)K(x, t, G(t)v(t))dt$$

where v is essentially bounded on I and $|v(t)| \leq |u|_{G^\infty}$ almost everywhere on I. We conclude that

$$\|G_{-1}(x) [u(x) - f(x)] \| \leq |u|_{G^\infty} \int_x^\infty g(t)dt$$

and the proof is complete. We remark that estimates of a similar nature were given by Love [7] in a more specific setting.

References

1. W.A. Coppel. Stability and asymptotic behavior of differential equations. D.C. Heath, Boston, 1965.

2. C. Corduneanu. " Problèmes globaux dans le théorie des équations intégrales de Vollterra." Ann. Mat. Pura Appl. 67 (1965) 349-363.

3. C. Corduneanu. "Some perturbation problems in the theory of integral equations." Math. Systems Theory. 1 (1967) 143-155.

4. A. Erdélyi. "The integral equations of asymptotic theory," Asymptotic Solutions of Differential Equations of Their Applications. John Wiley & Sons, New York, 1964, 211-229.

5. H.E. Gollwitzer. "Admissiblilty and integral operators" Math. Systems Theory. (to appear)

6. R. E. Langer. "Asymptotic theories for linear ordinary differential equations depending upon a parameter," SIAM J. Appl. Math. 7(1959) 298-305.

7. C.E. Love. "Singular integral equations of the Volterra type," Trans. Amer. Math. Soc. 15 (1914) 467-476.

8. J. L. Massera and J. J. Schaffer. Linear differential equations and function spaces. Academic Press, New York, 1966.

9. R.K. Miller. "Admissibility and nonlinear Volterra integral equations," Proc. Amer. Math. Soc. 25 (1970) 65-71.

10. R.K. Miller. Nonlinear Volterra Integral Equations. W.A. Benjamin, Inc., Menlo Park, California, 1971.

11. H.L. Turrittin. "Solvable related equations pertaining to turning point problems." Asymptotic Solutions of Differential Equations and Their Applications, John Wiley & Sons, New York, 1964, 27-52.

Differential Inequalities and Boundary Problems for Functional-Differential Equations

L. J. Grimm and L. M. Hall

1. Introduction

In the articles [1,2], Grimm and Schmitt employed differential inequalities, as developed by Jackson and Schrader [5] to study solutions of boundary value problems for a class of functional differential equations; their results included existence and uniqueness theorems together with *a priori* estimates on solutions. Recently, Kovač and coworkers (see, for instance, [6]) have developed an iteration scheme to improve such estimates. In this note we extend these results by a similar iteration. Further extensions are possible. For example, if the deviating arguments which are involved are solution-dependent, analogous theorems of existence, uniqueness, and inclusion will hold, but the estimates are not improved by iteration [2,4] . Higher order equations and systems of equations can be treated similarly [4,6]. Differential inequalities can also be used to provide comparision theorems, as in [3].

2. Existence and Uniqueness

Consider the boundary-value problem (BVP)

(2.1) $$Ly \equiv y''(t) + p(t)y'(t) + q(t)y(t) = f(t, y(t), y(g(t)), y'(t)) \equiv f[t, y]$$

on $I = (0,1)$, with the boundary conditions

(2.2) $$y(t) = \begin{cases} \phi_1(t), & t \in E_1 = [\alpha, 0] \\ \phi_2(t), & t \in E_2 = [1, \beta] \end{cases}$$

Supported in part by National Science Foundation Grant GP 27628.

where $\alpha = \min\{0, \inf_{t \varepsilon I} g(t)\}$, $\beta = \max\{1, \sup_{t \varepsilon I} g(t)\}$

with

(2.3) $$\phi_1(0) = \phi_2(1) = 0.$$

Here $f(t,u,v,w)$, $g(t)$, $\phi_1(t)$ and $\phi_2(t)$ are continuous functions on their respective domains.

Let $G(t,\tau)$ be the Green's function for the problem $Lu = 0$, $u(0) = u(1) = 0$. We assume throughout that $G(t,\tau) \leq 0$ on $0 \leq t, \tau \leq 1$. Sufficient conditions for this are given by Ževlakov and Pak [7, 8]. Let $G = \sup_{t \varepsilon I} \int_0^1 |G(t,\tau)|d\tau$ and $H = \sup_{t \varepsilon I} \int_0^1 |\frac{\partial G}{\partial t}(t,\tau)|\, d\tau$, and set $K = 2(G+H)$, and $E = E_1 \cup E_2$.

Existence and uniqueness of the solution of the problem (2.1) - (2.2) can be proved easily by fixed point techniques, see [2]. However, in the proof of the following theorem, which extends a result of Kovač [6], estimates are obtained which will be used later.

Theorem 1. Let $f(t,u,v,w)$ be bounded together with its partial derivatives f_2, f_3, f_4,

$$|f(t,u,v,w)| \leq B_0, \quad |f_2|, |f_3|, |f_4| \leq P,$$

when B_0 and P are constants, in the region

$$D = \{(t,u,v,w): 0 \leq t \leq 1, \ |u| \leq B_0 G, \ |v| \leq B_0 G, \ |w| < \infty\};$$

suppose that $P < \frac{1}{K}$. Let $|\phi_1(t)| \leq B_0 G$, $|\phi_2(t)| \leq B_0 G$. Then the problem (2.1) - (2.2) has a unique solution $y(t)$.

<u>Proof.</u> Let $z_1(t) \varepsilon$ $C(E \cup I)$ be a function satisfying

$z_1(t) = \phi_1(t)$, $t \varepsilon E_1$, $z_1(t) = \phi_2(t)$, $t \varepsilon E_2$, with $z_1(t) \varepsilon C^2(I)$. For $t \varepsilon I$, set

(2.4) $\alpha_1(t) = Lz_1 - f[t, z_1]$ and for $t \varepsilon E \cup I$,

(2.5) $z_{n+1}(t) = z_n(t) - \sigma_n(t)$, $n = 1, 2, \ldots$,

where $\sigma_n(t) \varepsilon C(E \cup I)$, $\sigma_n(t) \equiv 0$ on E,

(2.6) $L\sigma_n(t) = \alpha_n(t)$, $t \varepsilon I$, where the defect α_n is defined by

(2.7) $\alpha_n(t) = Lz_n - f[t, z_n]$, $t \varepsilon I$.

It is easily seen that $|z_n(t)| \leq B_0G$, $n = 1, 2, \ldots$, and that the iteration scheme (2.4) - (2.7) is equivalent to the iteration

(2.8) $$Lz_{n+1} = f[z_n],$$ since

(2.9) $\sigma_n(t) = \int_0^1 G(t, \tau)\, \alpha_n(t) d\tau$, $t \varepsilon I$, and thus (2.8) yields

(2.10) $z_{n+1}(t) = \int_0^1 G(t, \tau)\, f[\tau, z_n]\, d\tau$, $t \varepsilon I$.

We now show that $\lim_{n \to \infty} z_n$ exists and is a solution of (2.1)-(2.2).

For $t \varepsilon I$,

$$\alpha_{n+1}(t) = f[t, z_n] - f[t, z_{n+1}].$$

Hence

(2.11) $|\alpha_{n+1}(t)| \leq P[\, |\sigma_n(t)| + |\sigma_n(g(t))| + |\sigma'_n(t)|\,]$.

Let $M_0 = \max_{t \varepsilon I} |\alpha_1(t)|$. Then

$$|\sigma_1(t)| \leq M_0G, \quad |\sigma'_1(t)| \leq M_0H.$$

These estimates together with (2.11) yield

$$|\alpha_2(t)| \leq 2PM_0G + PM_0H = PM_0(2G + H) \leq PM_0K.$$

Continuing, we obtain

$$|\sigma_2(t)| \leq GPM_0K$$

$$|\sigma_2'(t)| \leq HPM_0K$$

and by induction

$$|\sigma_n(t)| \leq GM_0(PK)^{n-1}$$

$$|\alpha_n(t)| \leq M_0(PK)^{n-1} .$$

Hence if $PK < 1$, the series

$$z_1(t) + \sum_{n=1}^{\infty} [z_{n+1}(t) - z_n(t)]$$

and

$$Lz_1(t) + \sum_{n=1}^{\infty} L[z_{n+1}(t) - z_n(t)]$$

converge absolutely and uniformly on $E \cup I$; their respective sums $y_0(t)$ and $y_1(t)$ exist, are continuous, and y_0 satisfies the boundary conditions (2.2). Further, $y_1 = Ly_0$. To show that $Ly_0 = f$, we rewrite (2.10) as

$$\text{(2.12)} \qquad z_n(t) = \int_0^1 G(t,\tau) \{ \alpha_n(\tau) + f[\tau, z_n] \} d\tau .$$

By uniform convergence, we obtain

$$y_0(t) = \int_0^1 G(t,\tau) f[\tau, y_0] d\tau ,$$ i.e., $y_0(t)$ satisfies (2.1) - (2.2).

To show uniqueness, suppose there exists another solution y(t) of (2.1) - (2.2). Then

$$y(t) - y_0(t) = \int_0^1 G(t,\tau)\,\{f[\tau, y] - f[\tau, y_0]\}\,d\tau,$$

thus

$$|y'(t) - y_0'(t)| \leq \int_0^1 |\frac{\partial G}{\partial t}(t,\tau)|P\,\{\,|y(\tau) - y_0(\tau)| + |y(g(\tau)) - y_0(g(\tau))|$$

$$+ |y'(\tau) - y_0'(\tau)|\,\}\,d\tau .$$

For each function $\Psi(t)$ continuous on $E \cup I$, set $\|\Psi\| = \max_{E \cup I} |\Psi(t)|$, and set $x(t) = y(t) - y_0(t)$. By definition $x(t) \equiv 0$ on E.

Hence

$$\|x\| + \|x'\| \leq 2GP\,\|x\| + 2HP\,\|x\| + GP\,\|x\| + HP\,\|x'\|$$

$$\leq 2P(G + H)\,\{\,\|x\| + \|x'\|\,\}.$$ Thus

$\|x\| \equiv 0$, since $2P(G + H) < 1$ by hypothesis, and the proof is complete.

3. Inclusion Results

The iteration scheme developed above can be modified to yield monotonically convergent approximants to the solution of (2.1) - (2.2), provided that $f(t, u, v, w)$ is independent of w. We shall make this assumption, in addition to the hypotheses of Theorem 1, throughout this section. Let $B_1, B_2 < \infty$ be constants such that

(3.1) $\quad f_2(t, u, v) < B_1, \quad f_3(t, u, v) < B_2$

in the region D. Suppose that there exist functions $w_1(t)$ and $v_1(t)$ continuous on $E \cup I$, in class $C^2(I)$, with $|w_1|, |v_1| \leq B_0 G$, which satisfy the boundary conditions (2.2), and such that

(3.2) $\quad Lw_1 - f[t, w_1] - A_1(t) \equiv \gamma_1(t) \geq 0,$

(3.3) $\quad Lv_1 - f[t, v_1] + A_1(t) \equiv \beta_1(t) \leq 0,$

where

$$A_1(t) = [\,|B_1| + B_1][v_1(t) - w_1(t)] + [\,|B_2| + B_2][v_1(g(t) - w_1(g(t))\,].$$

Functions satisfying inequalities of the form (3.2) and (3.3) together with the boundary conditions (2.2) will be called, respectively, l-functions and u-functions for the BVP (2.1)-(2.2). Since $A_1(t) \geq 0$, these will be lower and upper solutions for this BVP, see [2].

We construct sequences of functions $\{w_n(t)\}$, $\{v_n(t)\}$, proceeding as in the construction of $\{z_n(t)\}$ in Theorem 1 :

(3.4) $$v_{n+1}(t) = v_n(t) - \omega_n(t),$$

where

(3.5) $$L\omega_n = \beta_n(t),\ t \in I,$$

$$\omega_n(t) \equiv 0,\ t \in E,$$

and

$$\beta_n(t) = Lv_n - f[t, v_n] + A_n(t)\ ;$$

(3.6) $$w_{n+1}(t) = w_n(t) - \rho_n(t),$$

where

$$L\rho_n = \gamma_n(t),\ t \in I,$$

$$\rho_n(t) \equiv 0,\ t \in E,$$

and

$$\gamma_n(t) = Lw_n - f[t, w_n] - A_n(t),$$

where

$$A_n(t) = [\,|B_1| + B_1][v_n(t) - w_n(t)] + [\,|B_2| + B_2][v_n(g(t)) - w_n(g(t))\,].$$

Note that if $B_1, B_2 \leq 0$, then $A_n(t) \equiv 0$ for all n.

Lemma. If there exist functions $w_1(t)$, $v_1(t)$ which are respectively l- and u-functions for the BVP (2.1) - (2.2) with $|w_1(t)|$, $|v_1(t)| \leq B_0 G$, and if the sequences $\{w_n(t)\}$, $\{v_n(t)\}$ are determined as above, then for all $n \geq 1$, $w_n(t) \leq w_{n+1}(t) \leq v_{n+1} \leq v_n(t)$.

Proof. Let $u_n(t) = v_n(t) - w_n(t)$. Equations (3.2) and (3.3) imply that

$$Lu_1 = \beta_1 - \gamma_1 + f[t, v_1] - f[t, w_1] - 2A_1(t).$$

By the mean-value theorem,

$$(3.7)\quad Lu_1 + [2|B_1| + 2B_1 - \tilde{f}_2]\, u_1(t) + [2|B_2| + 2B_2 - \tilde{f}_3]\, u_1(g(t))$$

$$- \beta_1(t) + \gamma_1(t) = 0,$$

(here $\tilde{f}_i$ denotes the value of the function f_i at some point in D) which is of the form

$$Lu_1 - h(t, u_1(t), u_1(g(t))) = 0,$$

where $h_2 \leq 0$, $h_3 \leq 0$. We shall show that $u_1(t) \geq 0$ and hence, that $w_1(t) \leq v_1(t)$. We form a sequence $\{Y_n(t)\}$ as follows: set

$$\Psi_n(t) = LY_n - h[t, Y_n], \text{ where}$$

$$(3.8)\quad Y_{n+1}(t) = Y_n(t) - \tau_n(t), \quad n = 1, 2, \ldots,$$

$$(3.9)\quad L\tau_n = \Psi_n,$$

$$\tau_n(t) \equiv 0, \ t \in E, \text{ and}$$

$$Y_1(t) \equiv 0.$$

Again by the mean-value theorem,

$$(3.10)\quad \Psi_{n+1}(t) = \tilde{h}_2\, \tau_n(t) + \tilde{h}_3\, \tau_n(g(t)).$$

Since $G(t,\tau) \leq 0$, (3.9) implies that $\tau_1(t) \leq 0$, and hence from (3.8), that $Y_1(t) \leq Y_2(t)$. Also, from (3.10), $\Psi_2(t) \geq 0$, hence $\tau_2(t) \leq 0$. By induction $\Psi_n(t) \geq 0$, $\tau_n(t) \leq 0$, $n = 1, 2, \ldots$. Thus $\{Y_n(t)\}$ is a monotonic increasing sequence which converges to the solution $u_1(t)$ of (3.7) which satisfies the homogeneous boundary conditions $\phi_1(t) = \phi_2(t) \equiv 0$ and thus $u_1(t) \geq 0$. Therefore $w_1(t) \leq v_1(t)$.

The rules for constructing $\{w_n(t)\}$ and $\{v_n(t)\}$ are equivalent to

$$Lw_{n+1} = f[t, w_n] + A_n(t), \tag{3.11}$$

$$Lv_{n+1} = f[t, v_n] - A_n(t). \tag{3.12}$$

Since $u_1(t) \geq 0$, (3.11) and (3.12) yield, for $n = 1$, $Lu_2 \leq 0$; thus $u_2(t) \geq 0$, hence $v_n(t) \geq w_n(t)$, $n = 1, 2, \ldots$. The monotonicity of the sequences $\{v_n(t)\}$, $\{w_n(t)\}$ follows as in the case of the $\{Y_n(t)\}$, if we use

$$\begin{aligned} \gamma_{n+1}(t) = {} & [-|B_1| - B_1 + \tilde{f}_2]\, \rho_n(t) + [-|B_2| - B_2 + \tilde{f}_3]\, \rho_n(g(t)) \\ & + [\,|B_1| + B_1]\, \omega_n(t) + [\,|B_2| + B_2]\, \omega_n(g(t)) \end{aligned} \tag{3.13}$$

and

$$\begin{aligned} \beta_{n+1}(t) = {} & [-|B_1| - B_1 + \tilde{f}_2]\, \omega_n(t) + [-|B_2| - B_2 + \tilde{f}_3]\, \omega_n(g(t)) \\ & + [\,|B_1| + B_1]\, \rho_n(t) + [\,|B_2| + B_2]\, \rho_n(g(t)). \end{aligned} \tag{3.14}$$

We find as in the first part of the proof that $\gamma_2(t) \geq 0$, $\beta_2(t) \leq 0$, hence $\rho_2(t) \leq 0$, $\omega_2(t) \geq 0$; an induction completes the proof.

<u>Theorem 2.</u> <u>Let the hypotheses of the above lemma be satisfied.</u>

<u>Let</u> $R = \max \{ \sup_D |-2|B_1| - 2B_1 + f_2|,\ \sup_D |-2|B_2| - 2B_2 + f_3| \}$

and define $\{w_n(t)\}$, $\{v_n(t)\}$ as in the lemma.

If $R < \frac{1}{4G}$, then $\lim_{n \to \infty} w_n(t) = \lim_{n \to \infty} v_n(t) = y(t)$ exists and is the unique solution of the BVP (2.1) - (2.2).

Proof. Uniqueness follows from Theorem 1. Using the notation of the lemma, set $M = \max \{ \sup_{t \varepsilon I} |\gamma_1(t)|, \sup_{t \varepsilon I} |\beta_1(t)| \}$.

From (3.5) with $n = 1$ it follows that

$$|\omega_1(t)| \leq MG; \text{ similarly}$$

$$|\rho_1(t)| \leq MG.$$

Equations (3.13) and (3.14) imply that

$$|\beta_2(t)| \leq 4RMG,$$

$$|\gamma_2(t)| \leq 4RMG, \text{ and by induction}$$

$$|\gamma_n(t)| \leq M(4RG)^{n-1},$$

$$|\beta_n(t)| \leq M(4RG)^{n-1}, \text{ and}$$

(3.15)
$$|\rho_n(t)| \leq MG(4RG)^{n-1}$$

$$|\omega_n(t)| \leq MG(4RG)^{n-1}.$$

Thus if $R < \frac{1}{4G}$, the series

$$w_1(t) + \sum_{n=1}^{\infty} [w_{n+1}(t) - w_n(t)],$$

$$v_1(t) + \sum_{n=1}^{\infty} [v_{n+1}(t) - v_n(t)],$$

$$Lw_1(t) + \sum_{n=1}^{\infty} [L(w_{n+1}(t) - w_n(t))],$$

$$Lv_1(t) + \sum_{n=1}^{\infty} [L(v_{n+1}(t) - v_n(t))]$$

converge absolutely and uniformly to the functions $w(t)$, $v(t)$, $\hat{w}(t)$, $\hat{v}(t)$ respectively, and as before, $Lw = \hat{w}$, $Lv = \hat{v}$. From the Lemma, $v_n(t) - w_n(t) \geq 0$, and thus $A_n(t) \geq 0$, $n = 1, 2, \ldots$. We now show that $\lim_{n \to \infty} \theta_n(t) = 0$, where $\theta_n(t) = v_n(t) - w_n(t)$. The function $\theta_n(t)$ is a solution of

$$L\theta_n = f[v_n] - f[w_n] + \beta_n - \gamma_n - 2A_n ,$$

satisfying homogeneous boundary conditions. We can write, using the mean-value theorem,

$$\begin{aligned} L\theta_n &= \tilde{f}_2 \theta_n(t) + \tilde{f}_3 \theta_n(g(t)) - 2(|B_1| + B_1)\, \theta_n(t) \\ &\qquad - 2(|B_2| + B_2)\, \theta_n(g(t)) + \beta_n - \gamma_n \\ &= [\tilde{f}_2 - 2(|B_1| + B_1)]\, \theta_n(t) + [\tilde{f}_3 - 2(|B_2| + B_2)]\, \theta_n(g(t)) \\ &\qquad + \beta_n - \gamma_n . \end{aligned}$$

Set $\|\theta_n\| = \sup_{t \in I} |\theta_n(t)|$.

$$\begin{aligned} |\theta_n(t)| &\leq \int_0^1 |\tilde{f}_2 - 2(|B_1| + B_1)|\, |G(t,\tau)|\, \|\theta_n\|\, d\tau \\ &\quad + \int_0^1 |\tilde{f}_3 - 2(|B_2| + B_2)|\, |G(t,\tau)|\, \|\theta_n\|\, d\tau \\ &\quad + \int_0^1 |G(t,\tau)|\, |\beta_n(\tau) - \gamma_n(\tau)|\, d\tau \\ &\leq 2RG \|\theta_n\| + T_n(t), \end{aligned}$$

where $T_n(t) \to 0$ uniformly on $[0,1]$ as $n \to \infty$. Hence $\theta_n \to 0$ uniformly in t as $n \to \infty$, and $\lim_{n \to \infty} v_n(t) = \lim_{n \to \infty} w_n(t)$, uniformly in t on $[0,1]$.

Denoting this limit by $y(t)$, we obtain finally from (3.11),

$$Ly = f[t, y]. \tag{3.16}$$

It is clear that y satisfies the boundary conditions (2.2), and the proof is complete.

Corollary. The functions $w_n(t)$, $v_n(t)$ determined as in the proof of the Lemma, satisfy

$$w_n(t) \leq y(t) \leq v_n(t).$$

4. A Numerical Example

The equation

$$y''(x) = -y(x) - y(x - \tfrac{1}{2})$$

with boundary conditions

$$y(x) = -x, x \in [-\tfrac{1}{2}, 0],$$

$$y(0) = y(1) = 0,$$

together with the functions

$$w_1(x) = \begin{cases} -x, x \in [-\tfrac{1}{2}, 0] \\ 0, x \in I \end{cases}$$

and

$$v_1(x) = \begin{cases} -x, x \in [-\tfrac{1}{2}, 0] \\ x - x^2, x \in I \end{cases}$$

satisfy the conditions of Theorem 2. The iteration scheme defined by

$$w_{n+1}(x) = \int_0^1 G(t,\tau)\, f[\tau, w_n]\, d\tau$$

$$v_{n+1}(x) = \int_0^1 G(t,\tau)\, f[\tau, v_n]\, d\tau$$

yields the following results. (As an example of the notation in the table, 0.4573D - 03 means 0.4573×10^{-3}).

Table I

t	$w_5(t)$	$v_5(t)$	$v_5(t)-w_5(t)$
0.000	0.0000D 00	0.0000D 00	0.0000D 00
0.125	0.2257D-02	0.2257D-02	0.1400D-09
0.250	0.2777D-02	0.2777D-02	0.2732D-08
0.375	0.1780D-02	0.1780D-02	0.9102D-08
0.500	0.0000D 00	0.1312D-07	0.1312D-07
0.625	0.3270D-04	0.3271D-04	0.1734D-07
0.750	0.3224D-04	0.3225D-04	0.1545D-07
0.875	0.1203D-04	0.1204D-04	0.8129D-08
1.000	0.0000D 00	0.0000D 00	0.0000D 00

Table II

t	$w_{20}(t)$	$v_{20}(t)$	$v_{20}(t) - w_{20}(t)$
0.000	0.0000D 00	0.0000D 00	0.0000D 00
0.125	0.2257D-02	0.2257D-02	0.4337D-18
0.250	0.2777D-02	0.2777D-02	0.0000D 00
0.375	0.1780D-02	0.1780D-02	0.0000D 00
0.500	0.0000D 00	0.6585D-35	0.6585D-35
0.625	0.3270D-04	0.3270D-04	0.3388D-20
0.750	0.3224D-04	0.3224D-04	0.0000D 00
0.875	0.1203D-04	0.1203D-04	0.0000D 00
1.000	0.0000D 00	0.0000D 00	0.0000D 00

References

1. L. J. Grimm and K. Schmitt, "Boundary value problems for delay differential equations," Bull. Amer. Math Soc., 74(1968), 997-1000.

2. L. J. Grimm and K. Schmitt, "Boundary value problems for differential equations with deviating arguments," Aequationes Math 3(1969), 321-322 ; 4 (1970), 176 - 190.

3. G. B. Gustafson and K. Schmitt, "Nonzero solutions of boundary value problems for second order ordinary and delay differential equations," J. Differential Equations 12 (1972) 129-147.

4. L. M. Hall, Inclusion Theorems for Boundary Value Problems for Delay Differential Equations, M.S. thesis, University of Missouri-Rolla, 1971.

5. L. K. Jackson and K. W. Schrader, "Comparison theorems for nonlinear differential equations," J. Differential Equation, 3 (1967), 248-255.

6. Ju. I. Kovač, "On a boundary problem for nonlinear systems of ordinary differential equations of higher order," Mat. Fiz. 6 (1969), 107-122 (Russian)

7. S. A. Pak, "On a priori bounds for solutions of boundary problems for second order ordinary differential equations," Diff. Urav. 3 (1967), 890-897 (Russian)

8. G. N. Ževlakov and S.A. Pak, "Conditions for negativity of the Green's function for Sturm-Liouville problems for linear second order ordinary differential equation," Diff. Urav. 5 (1969), 1114- 1119. (Russian)

Singularly Perturbed Boundary Value Problems Revisited

W. A. Harris, Jr.

1. Introduction

The author was introduced to singular perturbation problems at the Ph. D. thesis defence of G. Chapin [3], a student of H. L. Turrittin. This encounter led to a lasting interest and envolvement with such problems [4, 5, 6] in addition ot a deeper understanding of the author's own thesis, written under the direction of H. L. Turrittin.

Singular perturbation theory finds wide applicability and continues to be an area of active research activity. It seems appropriate at this Symposium to give a modern and simplified treatment of a singularly perturbed two point boundary value problem.

Consider the boundary value problem

$$(1.1)\qquad \begin{cases} y' = A(t)y + B(t)z \\ \varepsilon z' = C(t)y + D(t)z \end{cases}$$

$$(1.2)\qquad M(\varepsilon)\begin{pmatrix} y(0,\varepsilon) \\ z(0,\varepsilon)\end{pmatrix} + N(\varepsilon)\begin{pmatrix} y(1,\varepsilon) \\ z(1,\varepsilon)\end{pmatrix} = \begin{pmatrix} c_1(\varepsilon) \\ c_2(\varepsilon)\end{pmatrix}$$

on the interral $0 \le t \le 1$, where y, c_1 and z, c_2 are real m-dimensional and n-dimensional vectors respectively, and A, B, C, D, M, N are compatible matrices with appropriate orders.

Supported in part by the United States Army under Contract DA-ARO-D-31-124-71-G176

We are concerned with the existence of a solution to this boundary value problem,

$$\begin{pmatrix} y(t,\varepsilon) \\ z(t,\varepsilon) \end{pmatrix} ;$$

the limiting value of the solution as $\varepsilon \to 0+$,

$$\lim_{\varepsilon \to 0+} \begin{pmatrix} y(t,\varepsilon) \\ z(t,\varepsilon) \end{pmatrix} = \begin{pmatrix} \hat{y}(t) \\ \hat{z}(t) \end{pmatrix} ;$$

and the characterization of the limiting value of this solution in terms of an appropriate <u>degenerate</u> boundary value problem,

$$(1.3) \qquad \begin{cases} \hat{y}' = A(t)\hat{y} + B(t)\hat{z} \\ 0 = C(t)\hat{y} + D(t)\hat{z} \end{cases}$$

$$(1.4) \qquad \hat{M} \begin{pmatrix} \hat{y}(0) \\ \hat{z}(0) \end{pmatrix} + \hat{N} \begin{pmatrix} \hat{y}(1) \\ \hat{z}(1) \end{pmatrix} = \begin{pmatrix} \hat{c}_1 \\ \hat{c}_2 \end{pmatrix} .$$

2. A Canonical Differential System

We begin by transforming the linear differential system (1.1) into canonical form under

<u>Hypothesis I</u>: A, B, C, D are continuous and the eigenvalues of D have non-zero real part for $0 \le t \le 1$.

A straight forward calculation shows that the change of variables

$$(2.1) \qquad \begin{pmatrix} y \\ z \end{pmatrix} = \begin{pmatrix} I_m & -\varepsilon S \\ -T & I_n + \varepsilon TS \end{pmatrix} \begin{pmatrix} v \\ w \end{pmatrix} \equiv H(t,\varepsilon) \begin{pmatrix} v \\ w \end{pmatrix}$$

transforms the linear differential system (1.1) into the linear differential system

$$(2.2) \quad \begin{cases} v' = [A(t) - B(t)\, T(t,\varepsilon)]\, v \\ \varepsilon w' = [D(t) + \varepsilon T(t,\varepsilon) B(t)]\, w \end{cases}$$

provided that T and S are bounded solutions of the differential systems

$$(2.4) \quad \varepsilon T' = D(t)T - \varepsilon TA(t) + \varepsilon TB(t)T - C(t)$$

$$(2.5) \quad \varepsilon S' = \varepsilon [A(t) - B(t)T(t,\varepsilon)]\, S - S\, [D(t) + \varepsilon T(t,\varepsilon)B(t)] - B(t)$$

Clearly, if T, S exist, then the matrix $H(t,\varepsilon)$ is nonsingular for sufficiently small ε.

The transformation (2.1) will also affect the boundary conditions (1.2) which become

$$(2.6) \quad \tilde{M}(\varepsilon) \begin{pmatrix} v(0,\varepsilon) \\ w(0,\varepsilon) \end{pmatrix} + \tilde{N}(\varepsilon) \begin{pmatrix} v(1,\varepsilon) \\ w(1,\varepsilon) \end{pmatrix} = \begin{pmatrix} c_1(\varepsilon) \\ c_2(\varepsilon) \end{pmatrix},$$

where

$$\tilde{M}(\varepsilon) = M(\varepsilon)\, H(0,\varepsilon), \quad \tilde{N}(\varepsilon) = N(\varepsilon)\, H(1,\varepsilon).$$

Thus it is important that $\lim_{\varepsilon \to 0+} H(t,\varepsilon)$ be determined for all values of t, $0 \le t \le 1$. We will show that T and S may be determined so that

$$\lim_{\varepsilon \to 0+} T(t,\varepsilon) = D^{-1}(t)\, C(t), \qquad 0 \le t \le 1$$

$$\lim_{\varepsilon \to 0+} S(t,\varepsilon) = -B(t) D^{-1}(t), \qquad 0 \le t \le 1$$

and hence the limiting value of $\tilde{M}(\varepsilon)$ and $\tilde{N}(\varepsilon)$ can be characterized in terms of the limiting values of $M(\varepsilon)$ and $N(\varepsilon)$.

Lemma. The systems of differential equations

$$\varepsilon T' = D(t)\,T - TA(t) + \varepsilon TB(t)\,T - C(t)$$

$$\varepsilon S' = \varepsilon[A(t) - B(t)T(t,\varepsilon)]\,S - S[D(t) + \varepsilon\, T(t,\varepsilon)\,B(t)] - B(t)$$

have solutions $T(t,\varepsilon)$, $S(t,\varepsilon)$ which are uniformly bounded for $0 \le t \le 1$ and $0 < \varepsilon \le \varepsilon_0$, ε_0 sufficiently small. Further,

$$\lim_{\varepsilon \to 0+} T(t,\varepsilon) = D^{-1}(t)\,C(t),$$

$$\lim_{\varepsilon \to 0+} S(t,\varepsilon) = -B(t)\,D^{-1}(t), \qquad 0 \le t \le 1.$$

Proof. We confine our attention to the differential system (2.4). Let $Z(t,\varepsilon)$ be a fundamental matrix for the linear differential system $\varepsilon Z' = D(t)\,Z$ and let $Y(t)$ be a fundamental matrix for the linear differential system $Y' = A(t)Y$. The general solution of the linear differential system $\varepsilon T' = D(t)T - \varepsilon\, TA(t)$ is of the form

$$T(t,\varepsilon) = Z(t,\varepsilon)\,E(\varepsilon)\,Y^{-1}(t).$$

If we consider $\varepsilon\, TB(t)T - C(t) \equiv G$, as a nonhomogeneous term, we may apply the method of variation of parameters to determine a solution of the differential system $\varepsilon T' = D(t)T - \varepsilon\, T\,A(t) + G$ in the form $T = ZEY^{-1}$ where

$$\varepsilon E' = Z^{-1}(t,\varepsilon)\,GY(t).$$

Hence we may write the differential system (2.4) in the form

$$T(t,\varepsilon) = \int^{t} Z(t,\varepsilon)\,Z^{-1}(s,\varepsilon)\,[T(s,\varepsilon)\,B(s)T(s,\varepsilon) - \varepsilon^{-1}\,C(s)]\;Y(s)Y^{-1}(t)ds,$$

where the path of integration is as yet unspecified.

Our assumption on D implies that D^{-1} exists and that D has the constant number $p(0 \le p \le n)$ of eigenvalues with negative real part for $0 \le t \le 1$.

Hence, if

$$P = \begin{pmatrix} I_p & 0 \\ 0 & 0 \end{pmatrix}$$

we have the fundamental inequalities

$$(2.7)\quad \begin{cases} \| Z(t,\varepsilon)\, P Z^{-1}(s,\varepsilon) \| \leq L \exp\{\frac{-\mu(t-s)}{\varepsilon}\} \text{ for } 1 \geq t \geq s \geq 0, \\ \| Z(t,\varepsilon)\, (I_n - P) Z^{-1}(s,\varepsilon) \| \leq L \exp\{\frac{-\mu(s-t)}{\varepsilon}\} \text{ for } 1 \geq s \geq t \geq 0 \\ \| Y(t)\, Y^{-1}(s) \| \leq \exp\{\sigma |t-s|\} \text{ for } 0 \leq t, s \leq 1 \end{cases}$$

where L, μ, σ are positive constants independent of ε.

We consider the integral equation $T = \mathcal{T} T$ where

$$\begin{aligned} \mathcal{T}T(t,\varepsilon) = & \int_0^t Z(t,\varepsilon)\, P Z^{-1}(s,\varepsilon)\, [T(s,\varepsilon)\; B(s) T(s,\varepsilon) - \varepsilon^{-1} C(s)]\, Y(s) Y^{-1}(t) ds \\ & + \int_1^t Z(t,\varepsilon)\, (I_n - P) Z^{-1}(s,\varepsilon)\, [T(s,\varepsilon)\, B(s) T(s,\varepsilon) - \varepsilon^{-1} C(s)]\, Y(s) Y^{-1}(t) ds \\ & + Z(t,\varepsilon)\, P Z^{-1}(0,\varepsilon)\, D^{-1}(0)\, C(0)\, Y(0) Y^{-1}(t) \\ & + Z(t,\varepsilon) (I_n - P) Z^{-1}(1,\varepsilon)\, D^{-1}(1)\, C(1)\, Y(1)\, Y^{-1}(t) \end{aligned} \quad (2.8)$$

Utilizing our fundamental inequalities (2.7), it follows by the contraction principle that the integral equation $T = \mathcal{T} T$ has a unique solution that can be obtained by successive approximation and that this solution is a solution of the differential system (2.4).

To show that $\lim_{\varepsilon \to 0+} T(t,\varepsilon) = D^{-1}(t)\, C(t)$ we use the identity

$$H(t) = -\int_0^t Z(t,\varepsilon)\, PZ^{-1}(s,\varepsilon)\, \varepsilon^{-1}D(s)H(t)Y(s)Y^{-1}(t)\, ds$$

$$- \int_1^t Z(t,\varepsilon)\, (I_n-P)Z^{-1}(s,\varepsilon)\varepsilon^{-1}D(s)H(t)Y(s)Y^{-1}(t)ds$$

$$+ \int_0^t Z(t,\varepsilon)\, PZ^{-1}(s,\varepsilon)H(t)A(s)Y(s)Y^{-1}(t)ds$$

$$+ \int_1^t Z(t,\varepsilon)(I_n-P)Z^{-1}(s,\varepsilon)H(t)A(s)Y(s)Y^{-1}(t)ds$$

$$+ Z(t,\varepsilon)PZ^{-1}(0,\varepsilon)H(t)Y(0)Y^{-1}(t) + Z(t,\varepsilon)(I_n-P)Z^{-1}(1,\varepsilon)H(t)Y(1)Y^{-1}(t).$$

Setting $H(t) = D^{-1}(t)\, C(t)$ and utilizing the integral equation for T and the fundamental inequalities (2.7) we obtain the desired result.

3. A Related Boundary Value Problem

The change of variable (2.1) has transformed the original boundary value problem (1.1) (1.2) into the related boundary value problem (2.2), (2.6) which is more tractable.

We seek a solution to this boundary value problem in the form

$$(3.1)\quad \begin{pmatrix} v(t,\varepsilon) \\ w(t,\varepsilon) \end{pmatrix} = \begin{pmatrix} V(t,\varepsilon) & 0 \\ 0 & W(t,\varepsilon)PW^{-1}(0,\varepsilon)+W(t,\varepsilon)(I_n-P)W^{-1}(1,\varepsilon) \end{pmatrix} \begin{pmatrix} \alpha_1(\varepsilon) \\ \alpha_2(\varepsilon) \end{pmatrix}$$

where V, W are fundamental matrices of the linear differential systems

$$V' = (A-BT)V$$

$$\varepsilon W' = (D+\varepsilon\, TB)W,$$

and α_1, α_2 are vectors to be determined, i.e.

$$(3.2)\qquad \Delta(\varepsilon)\begin{pmatrix}\alpha_1(\varepsilon)\\ \alpha_2(\varepsilon)\end{pmatrix} = \begin{pmatrix}c_1(\varepsilon)\\ c_2(\varepsilon)\end{pmatrix}$$

where

$$(3.3)\qquad \Delta(\varepsilon) = \tilde{M}(\varepsilon)\ \mathrm{diag}\,[V(0,\varepsilon), W(0,\varepsilon)PW^{-1}(0,\varepsilon) + W(0,\varepsilon)(I_n-P)W^{-1}(1,\varepsilon)]$$

$$+ \tilde{N}(\varepsilon)\ \mathrm{diag}\,[V(1,\varepsilon), W(1,\varepsilon)PW^{-1}(0,\varepsilon)+W(1,\varepsilon)(I_n-P)W^{-1}(1,\varepsilon)]\ .$$

If $\Delta^{-1}(\varepsilon)$ exists, then the solution of the boundary value problem is

$$(3.4)\qquad \begin{pmatrix}v(t,\varepsilon)\\ w(t,\varepsilon)\end{pmatrix} = \begin{pmatrix}V(t,\varepsilon & 0\\ 0 & W(t,\varepsilon)PW(0,\varepsilon)+W(t,\varepsilon)(I_n-P)W^{-1}(1,\varepsilon)\end{pmatrix}\Delta^{-1}(\varepsilon)\begin{pmatrix}c_1(\varepsilon)\\ c_2(\varepsilon)\end{pmatrix}.$$

Let us analyse $\Delta(\varepsilon)$. Since

$$W(1,\varepsilon)\ PW^{-1}(0,\varepsilon) = 0(\varepsilon),\quad W(0,\varepsilon)(I-P)W^{-1}(1,\varepsilon) = 0(\varepsilon)$$

as $\varepsilon \to 0+$, we may write

$$\Delta(\varepsilon) = (\tilde{M}_1(0)V(0) + \tilde{N}_1(0)V(1) : \ \tilde{M}_2(0)W(0)PW^{-1}(0)+\tilde{N}_2(0)W(1)(I_n-P)\ W^{-1}(1)) + 0(\varepsilon)$$

where $V(t) = \lim_{\varepsilon \to 0+} V(t,\varepsilon)$, $W(t) = \lim_{\varepsilon \to 0+} W(t,\varepsilon)$, and the compatible partitioning

$$\tilde{M}(\varepsilon) = (\tilde{M}_1(\varepsilon) : \tilde{M}_2(\varepsilon))\ ,\quad \tilde{N}(\varepsilon) = (\tilde{N}_1(\varepsilon) : \tilde{N}_2(\varepsilon))$$

with

$$\tilde{M}_i(0) = \lim_{\varepsilon \to 0+} \tilde{M}_i(\varepsilon)\ ,\quad \tilde{N}_i(0) = \lim_{\varepsilon \to 0+} \tilde{N}_i(\varepsilon)\ .$$

Therefore, for all sufficiently small ε, $\Delta^{-1}(\varepsilon)$ will exist if we make

<u>Hypothesis II:</u> the matrix

$$\Delta(0) = (\tilde{M}_1(0)V(0) + \tilde{N}_1(0)V(1) \vdots \ \tilde{M}_2(0)W(0)PW^{-1}(0) + \tilde{N}_2(0)W(1)(I_n - P)W^{-1}(1))$$

is nonsingular.

In view of the fundamental inequalities (2.7) it follows that for $0 < t < 1$

$$\mathbf{x}(t) \equiv \begin{pmatrix} \mathbf{x}_1(t) \\ \mathbf{x}_2(t) \end{pmatrix} \equiv \lim_{\varepsilon \to 0+} \begin{pmatrix} v(t, \varepsilon) \\ w(t, \varepsilon) \end{pmatrix} = \begin{pmatrix} V(t) & 0 \\ 0 & 0 \end{pmatrix} \begin{pmatrix} \alpha_1(0) \\ \alpha_2(0) \end{pmatrix} = \begin{pmatrix} V(t)\alpha_1(0) \\ 0 \end{pmatrix},$$

that is $\mathbf{x}(t)$ satisfies the corresponding degenerate differential system of (2.2)

$$\mathbf{x}_1^1 = [A(t) - B(t)D^{-1}(t)C(t)]\,\mathbf{x}_1$$

$$0 = D(t)\,\mathbf{x}_2 .$$

Also $\mathbf{x}(t)$ satisfies the first m boundary conditions of

$$\Delta^{-1}(0) \{ \tilde{M}(0)\mathbf{x}(0) + \tilde{N}(0)\,\mathbf{x}(1) \} = \Delta^{-1}(0) \begin{pmatrix} c_1(0) \\ c_2(0) \end{pmatrix}$$

Returning to the original boundary value problem, we have the following

Theorem. Let Hypotheses I, II hold. For sufficiently small $\varepsilon > 0$ the boundary value problem

$$y' = A(t)y + B(t)z$$

$$\varepsilon z' = C(t)\,y + D(t)z$$

$$M(\varepsilon) \begin{pmatrix} y(0, \varepsilon) \\ z(0, \varepsilon) \end{pmatrix} + N(\varepsilon) \begin{pmatrix} y(1, \varepsilon) \\ z(1, \varepsilon) \end{pmatrix} = \begin{pmatrix} c_1(\varepsilon) \\ c_2(\varepsilon) \end{pmatrix}$$

has the solution

$$\begin{pmatrix} y(t,\varepsilon) \\ z(t,\varepsilon) \end{pmatrix} = \begin{pmatrix} I_m & -\varepsilon S(t,\varepsilon) \\ -T(t,\varepsilon) & I_n + T(t,\varepsilon) S(t,\varepsilon) \end{pmatrix} \begin{pmatrix} v(t,\varepsilon) \\ w(t,\varepsilon) \end{pmatrix}$$

for $0 \leq t \leq 1$. Further, for $0 < t < 1$

$$\lim_{\varepsilon \to 0+} \begin{pmatrix} y(t,\varepsilon) \\ z(t,\varepsilon) \end{pmatrix} = \begin{pmatrix} \hat{y}(t) \\ \hat{z}(t) \end{pmatrix}$$

where

$$\hat{y}' = A(t)\,\hat{y} + B(t)\,\hat{z}$$

$$0 = C(t)\,\hat{y} + D(t)\,\hat{z}$$

and

$$\begin{pmatrix} I_m & 0 \\ 0 & 0 \end{pmatrix} \Delta^{-1}(0) \left\{ M(0) \begin{pmatrix} \hat{y}(0) \\ \hat{z}(0) \end{pmatrix} + N(0) \begin{pmatrix} \hat{y}(1) \\ \hat{z}(1) \end{pmatrix} \right\}$$

$$= \begin{pmatrix} I_m & 0 \\ 0 & 0 \end{pmatrix} \Delta^{-1}(0) \begin{pmatrix} c_1(0) \\ c_2(0) \end{pmatrix},$$

or

$$\hat{M} \begin{pmatrix} \hat{y}(0) \\ \hat{z}(0) \end{pmatrix} + \hat{N} \begin{pmatrix} \hat{y}(1) \\ \hat{z}(1) \end{pmatrix} = \begin{pmatrix} \hat{c}_1 \\ \hat{c}_2 \end{pmatrix}.$$

4. Remarks

It is clear that more general boundary value problems can be handled in the same manner. However, if more than a first term approximation is desired, the coefficient matrices A, B, C, D, M, N will have to be more regular.

The essence of the method presented here is the use of solutions of nonlinear differential systems to obtain a canonical form for a given differential system. This technique is now a standard method, see Y. Sibuya [9], K. W. Chang [1, 2].

Hypotheses I is natural and reasonable. Hypotheses II is perhaps the simplest sufficient condition for establishing this type of result.

It the problem was originally cast as a singular pertubation problem consisting of an n th order differential equation with separated boundary conditions, $\Delta(\varepsilon)$ can be analyzed in a straight forward manner to give the usual cancellation laws, see W. Wasow [10], R.E. O'Malley [7, 8].

References

1. K. W. Chang. "Remarks on a certain hypothesis in singular pertubations", Proc. Amer. Math. Soc. 23(1969), 41-45.

2. K.W. Chang. " Singular perturbations of a general boundary value problem", (to appear).

3. G. G. Chapin, Jr. One and two point boundary value problems for ordinary differential equations containing a parameter, Ph.D. Thesis, University of Minnesota, Minneapolis, 1959.

4. W. A. Harris, Jr. "Singular perturbations of two-point boundary problems for systems of ordinary differential equations," Arch. Rat. Mech. Anal. 5 (1960), 212-225.

5. W. A. Harris, Jr. "Singular perturbations of two point boundary problems," J. Math. Mech. 11 (1962), 371-382.

6. W. A. Harris, Jr. "Equivalent classes of singular perturbation problems," Rend. Circ. Matem. di Palermo, 14 (1965), 1-15.

7. R. E. O'Malley, Jr. "Boundary value problems for linear systems of ordinary differential equations involving many small parameters," J. Math. Mech. 18 (1969), 835-855.

8. R. E. O'Malley, Jr. and J. B. Keller. "Loss of boundary conditions in the asymptotic solution of linear differential equations. II. Boundary value problems," Comm. Pure Appl. Math. 21 (1968), 263-270.

9. Y. Sibuya. "Sur réduction analytique d'an systeme d'equations différentielles ordinaires linéaires contenant un paramètre", J. Fac. Sci, Univ. Tokyo, Sec I. 7 (1968), 527-540.

10. W. Wasow. "On the asymptotic solution of boundary value problems for ordinary differential equations containing a parameter," J. Math. Phys. 23 (1944), 173-183.

Bounded Solutions of Nonlinear Equations at an Irregular Type Singularity

P. F. Hsieh

1. Introduction

An important task in the analytic theory of ordinary differential equations at a singular point is to reduce the given equations into a form as simple as possible, and then study the simplified equations. In the course of simplification, not only the formal reduction is to be found, also its analytic meaning should be investigated. Of course, it is desirable to prove the convergence of the formal reduction, but, when this is impossible, it is attempted to find an analytic reduction which admits the formal reduction as its asymptotic expansion in its domain of validity. If these are done, an analytic expression of the solutions of the original equations can be given as an analytic function of the solutions of the simplified equations. Many authors studied the analytic solutions of linear equations at an irregular singular point in this spirit (eg. see [4]). In this note, we shall study the analytic expression of bounded solutions of nonlinear differential equations at an irregular type singularity.

Consider a system of nonlinear equations of the form :

$$x^{\sigma+1}y' = f(x, y, z) \,, \quad xz' = g(x, y, z) \tag{1.1}$$

where x is a complex independent variable, σ is a positive integer, y and f are m-column vectors, z and g are n-column vectors, f and g are holomorphic in a neighborhood of $(0, 0, 0)$ and $f(0, 0, 0) = 0$, $g(0, 0, 0) = 0$. Furthermore, the Jacobian matrices $A = f_y(0, 0, 0)$ and $B = g_z(0, 0, 0)$ are non-singular. In this case, $x = 0$ is called an irregular type singularity of (1.1).

This research is partially supported by NSF Grant GP-14595. Part of this work is done while the author was on sabbatical leave at Naval Research Laboratory, Washington, D.C.

For an n-row vector $q = (q_1, \ldots, q_n)$ of non-negative integers q_k and an n-column vector z with elements $\{z_1, \ldots, z_n\}$, we denote

$$|q| = q_1 + \cdots + q_n ,$$

$$\|z\| = \max_{k=1}^{n} |z_k| ,$$

$$q \cdot z = q_1 z_1 + \cdots + q_n z_n ,$$

and

$$1_n(z) = \operatorname{diag}(z_1, \ldots, z_n).$$

We shall assume that (1.1) satisfies the following conditions:

I. $B = g_z(0,0,0) = 1_n(\mu)$, where $\mu = \operatorname{col}(\mu_1, \ldots, \mu_n)$, and

$$\operatorname{Re} \mu_k > 0, \ (k = 1, 2, \ldots, n).$$

II. For any $(1 + n)$-row vectors $(\ell, q_1, \ldots, q_n)$ of non-negative integers satisfying $\ell + |q| \geq 2$, we have

$$\mu_k \neq \ell + q \cdot \mu , \quad \text{for } k = 1, 2, \ldots, n.$$

III. Let $\nu_1, \nu_2, \ldots, \nu_m$ be eigenvalues of $A = f_y(0,0,0)$, then

$$\operatorname{Re} \nu_1 \geq \operatorname{Re} \nu_2 \geq \cdots \geq \operatorname{Re} \nu_m > 0.$$

Under these conditions, we shall find an analytic expression of the solutions of (1.1) which are bounded as x tends to 0 along positive real axis. Such solutions are called bounded solutions.

The same study has been done by M. Iwano [3] under two more restrictive conditions; namely

IV. The eigenvalues $\{\nu_j\}$ of A are mutually distinct.

V. For $p = (p_1, \ldots, p_m)$ of non-negative integers such that $|p| \geq 2$,

$$\nu_j \neq p \cdot \nu \text{ for } j = 1, 2, \ldots, m.$$

This study is a generalization of Iwano's work without conditions IV and V. However, we shall assume that the matrix $f_y(x, 0, z)$ can be diagonalized up to the x^σ term when it is expanded in power series of x, as will be seen in (3.4) below.

The system (1.1) will be reduced to the simplest form in three steps, each is stated as a theorem below. The proof of these theorems are lengthly and will appear elsewhere. In the proof of each of these theorems, two type of existence theorems are needed. These existence theorems are first proved by Iwano [3] by means of Tychonoff type fixed point theorem. Recently, the author [1] proved these by successive approximations method utilizing Iwano's scheme of integral equations involving improper contour integrations in the complex plane.

The author thanks Prof. M. Iwano for useful discussions.

2. Preliminary reduction

Let $\nu_1, \nu_2, \ldots, \nu_s$ be distinct eigenvalues of A with multiplicities $m_1, m_2, \ldots, m_s$ $(m_1 + m_2 + \cdots + m_s = m)$ respectively. Assume, without loss of generality, that A is in the following Jordan canonical form

$$(2.1) \qquad A = \sum_{j=1}^{s} \oplus (v_j 1_{m_j} + N_j)$$

where $\sum \oplus$ denotes direct sum, 1_m is the m by m identity matrix and N_j denotes an m_j by m_j nilpotent matrix with 0 everywhere except possibly 1 at subdiagonal elements.

We have the following

Theorem 1. *Assume that the conditions I are II and satisfied. Then there exists a transformation*

$$(2.2) \qquad y = P(x, \hat{z}) \hat{y} + \Phi(x, \hat{z}), \quad z = \Psi(x, \hat{z})$$

such that

(i) $P(x, \hat{z})$, $\Phi(x, \hat{z})$ and $\Psi(x, \hat{z})$ are m by m, m by 1 and n by 1 matrices, respectively, with their elements holomorphic in a domain of the form

$$(2.3)\qquad 0 < |x| < a, \quad \theta_1 < \arg x < \theta_1', \quad \|\hat{z}\| \leq c$$

where a and c are positive constants θ_1 and θ_1' are suitably chosen constants satisfying

$$(2.4)\qquad -\frac{3\pi}{2\sigma} \leq \theta_1 < 0 < \theta_1' \leq \frac{3\pi}{2\sigma} \;;$$

(ii) The system (1.1) is reduced to

$$(2.5)\qquad \begin{cases} x^{\sigma+1}\hat{y}' = F_0(x, \hat{z})\,\hat{y} + \sum\limits_{|p|\geq 2} \hat{y}^p F_p(x, \hat{z}) \\ x\,\hat{z}' = 1_n(\mu)\,\hat{z} + D(x, \hat{z})\,\hat{y} + \sum\limits_{|p|\geq 2} \hat{y}^p F_p(x, \hat{z}) \end{cases}$$

where F_0, F_p, D and G_p are m by m, m by 1, n by n and n by 1 matrices holomorphic in (2.3), and the right hand side of (2.5) are uniformly convergent in

$$(2.6)\qquad 0 < |x| < a, \quad \theta_1 < \arg x < \theta_1', \quad \|\hat{y}\| \leq b, \quad \|\hat{z}\| < c$$

where b is a positive constant.

Furthermore, $F_0(x, \hat{z})$ is in the block-diagonal form agreeing with that of A and

$$(2.7)\qquad F_0(0,0) = A, \quad D(0,0) = 0.$$

This theorem can be proved in two steps. First a transformation by means of a particular solution such as that found by M. Iwano [3] can be applied to reduce the equations to those with linear coefficients of y satisfying (2.7). Then, a block-diagonalization theorem proved by the author [2] can be applied to block-diagonalize $F_0(x, \hat{z})$.

3. Main reductions

Let the block-diagonal matrix $F_0(x, \hat{z})$ be denoted as the following

$$F_0(x, \hat{z}) = \sum_{j=1}^{s} \oplus F_{0j}(x, \hat{z}) \tag{3.1}$$

where $F_{0j}(x, \hat{z})$ are m_j by m_j matrices. Furthermore, put

$$F_{0j}(x, \hat{z}) = f_{j0}(\hat{z}) + f_{j1}(\hat{z})\, x + \cdots + f_{j\sigma}(\hat{z})\, x^{\sigma} + x^{\sigma+1} F_j^0(x, \hat{z}) \tag{3.2}$$

where $f_{jk}(\hat{z})$ $(j = 1, \ldots, s, \ k = 0, 1, \ldots, \sigma)$ are m_j by m_j matrices holomorphic in

$$\| \hat{z} \| \leq c \tag{3.3}$$

and $F_j^0(x, \hat{z})$ are m_j by m_j matrices holomorphic in (2.3). Assume that

$$f_{j0}(z) + f_{j1}(z)x + \cdots + f_{j,\sigma-1}(z)\, x^{\sigma-1} + f_{j\sigma}(0)\, x^{\sigma} = \lambda_j[x, z]\, 1_{m_j} \tag{3.4}$$

where $\lambda_j[x, z]$ is a scalar quantity. Then, we can prove

Theorem 2. *There exists a transformation*

$$\hat{y} = Y + \left(\sum_{j=1}^{s} \oplus R_j(x, Z) \right) Y, \quad \hat{z} = Z + x^{\sigma} S(x, Z)\, Y \tag{3.5}$$

such that

(i) *R_j and S are m_j by m_j $(j = 1, \ldots, s)$ and n by n matrices, respectively, holomorphic in*

(3.6) $$0 < |x| a_1, \quad \theta_2 < \arg x < \theta_2', \quad \|Z\| \leq c_1$$

<u>with</u> $0 < a_1 \leq a,\ 0 < c_1 \leq c,\ \theta_1 \leq \theta_2 < 0 < \theta_2' \leq \theta_1'$;

(ii) <u>The system</u> (2.5) <u>is reduced to</u>

(3.7) $$\begin{cases} x^{\sigma+1} Y' = \left(\sum_{j=1}^{s} \oplus \lambda_j[x, Z]\, 1_{m_j} \right) Y + \sum_{|p| \geq 2} Y^p \tilde{F}_p(x, Z) \\ xZ' = 1_n(\mu) Z + \sum_{|p| \geq 2} Y^p \tilde{G}_p(x, Z) \end{cases}$$

<u>where</u> $\tilde{F}_p$ <u>and</u> $\tilde{G}_p$ <u>are</u> m-<u>column and</u> n-<u>column vectors, respectively, holomorphic in</u> (3.6), <u>and the right hand side of</u> (3.7) <u>are uniformly convergent in</u>

(3.8) $$0 < |x| < a_1, \quad \theta_2 < \arg x < \theta'_2, \quad \|Y\| \leq b_1, \quad \|Z\| < c_1$$

<u>with</u> $0 < b_1 \leq b$.

In order to simplify further the equations, let R be the set of m-tuples of non-negative integers $p = (p_1, \ldots, p_m)$

(3.9) $$R = \{ p \mid |p| \geq 2 \text{ and there exists } j \text{ such that}$$

$$\nu_j = (p_1 + \cdots + p_{m_1}) \nu_1 + \cdots + (p_{m-m_s+1} + \cdots + p_m) \nu_s \} .$$

By assumption III, there are only finite number of m-tuples in R. In terms of this set, we can prove

<u>Theorem</u> 3. <u>There exists a transformation</u>

(3.10) $$Y = u + \sum_{|p| \geq 2} u^p A_p(x, v), \quad Z = v + x^\sigma \sum_{|p| \geq 2} u^p B_p(x, v)$$

such that

(i) A_p *and* B_p *are* m-*column and* n-*column vectors, respectively, holomorphic in*

(3.11) $$0 < |x| < a_2, \ \theta_3 < \arg x < \theta'_3, \ \|v\| \le c_2$$

with $0 < a_2 \le a_1, \ \theta_2 \le \theta_3 < 0 < \theta'_3 \le \theta'_2, \quad 0 < c_2 \le c_1$

and the right hand side converges uniformly in

(3.12) $0 < |x| < a_2, \ \theta_3 < \arg x < \theta'_3, \ \|u\| \le b_2, \ \|v\| \le c_2$

with $0 < b_2 \le b_1$;

(ii) *The system* (3.7) *is reduced to*

(3.13) $$\begin{cases} x^{\sigma+1}u' = \left(\sum_{j=1}^{s} \oplus \lambda_j[x,v]\, 1_{m_j} \right) u + \sum_{p \in R}' u^p K_p(x,v) \\ xv' = 1_n(\mu)\, v \end{cases}$$

where $\sum'$ *denotes summation over* p *in* R, K_p *are* m-*column vectors holomorphic in* (3.11).

Furthermore, for p *satisfying*

(3.14) $$\lambda_j[x,v] \equiv (p_1 + \cdots + p_{m_1})\lambda_1[x,v] + \cdots + (p_{m-m_s+1} + \cdots + p_m)\lambda_s[x,v],$$

(*for all* j = 1, 2, ..., s),

$K_p(x,v)$ *is a polynomial of* x *with degree* $\le \sigma$ *and coefficients holomorphic in*

(3.15) $$\|v\| \le c_2,$$

4. Analytic expression of bounded solutions

Assume that for all p in R, (3.14) is satisfied. Then all the quantities K_p are polynomials of x with degree $\leq \sigma$, and an analytic expression of the bounded solutions of (3.13) can be obtained by quadrature.

Let V(x) be a general solution of the second equations of (3.13); namely $V(x) = 1_n(x^{\mu})\, C$ where x^{μ} denotes the n-column vector with elements $\{x^{\mu_k}\}$ and C is an arbitrary n-column vector. By integration by parts, we have

$$\int \frac{\log x}{x^{s+1}}\, dx = -\frac{1}{s x^s}\left[\log x + \frac{1}{s}\right] \quad s \neq 0,$$

$$\int \frac{V(x)^q}{x^{s+1}}\, dx = \begin{cases} \dfrac{V(x)^q}{(q\cdot\mu - s)x^s}, & \text{if } q\cdot\mu \neq s \\ C^q \log x, & \text{if } q\cdot\mu = s, \end{cases}$$

$$\int \frac{V(x)^q (\log x)^k}{x^{s+1}}\, dx = \begin{cases} \dfrac{V(x)^q}{(q\cdot\mu - s)x^s}\left[\sum_{h=0}^{k} (-1)^h \dfrac{(\log x)^{k-h}}{(q\cdot\mu - s)^h}\right], & \text{if } q\cdot\mu \neq s \\ C^q \dfrac{(\log x)^{k+1}}{k+1}, & \text{if } q\cdot\mu = s. \end{cases}$$

Then, the analytic expression of a bounded solution of (3.13) when all K_p are polynomials in x with degree $\leq \sigma$ is $\{U(x, V(x)), V(x)\}$ where

$$U(x, v) = \text{col}\,(U_1(x, v),\ U_2(x, v),\ \ldots,\ U_s(x, v))$$

with $U_j(x,v)$ m_j-column vectors of the form

$$U_j(x,v) = e^{\Omega_j[x,v]} x^{\lambda_j(C)} \{ C_j + \psi_j(x\, ;\, C_{j+1}, \ldots, C_s, \log x, v)) \}.$$

Here $\Omega_j[x,v]$ is a polynomial of x^{-1} with degree $\leq \sigma$ and coefficients holomorphic in $\|v\| \leq c_2$, $\lambda_j(C)$ is a polynomial of C such that $\lambda_j(0) = f_{j\sigma}(0)$, C_j is an arbitrary constant m_j - column vector and ψ_j are m_j - column vector whose entries are polynomial of x^{-1} with degree $\leq \sigma$ and coefficients polynomials of C, $C_{j+1}, \ldots, C_s$, $\log x$ and holomorphic in v for $\|v\| \leq c_2$. Furthermore $\psi \equiv 0$.

References

1. P.F. Hsieh. "Successive approximations method for solutions of nonlinear differential equations at an irregular type singular points," Comment. Math. Univ. St. Pauli, 20 (1971), 27-53.

2. P. F. Hsieh. "Analytic simplification of a system of ordinary differential equations at an irregular type singularity," Comment. Math. Univ. St. Pauli 20 (1971), 55-82.

3. M. Iwano. "Analytic expressions for bounded solutions of non-linear ordinary differential equations with an irregular type singular point," Ann. Mat. Pura Appl. (4), 82 (1969), 189-256.

4. H. L. Turrittin. "Convergent solutions of ordinary linear homogeneous differential equations in the neighborhood of an irregular singular point," Acta Math. 93 (1955), 27-66.

On Meromorphic Solutions of the Difference Equation

$$y(x+1) = y(x) + 1 + \frac{\lambda}{y(x)}$$

Tosihusa Kimura

1. Introduction

In this note we shall be concerned with the difference equation

$$\text{(E)} \qquad y(x+1) = y(x) + 1 + \frac{\lambda}{y(x)} ,$$

where λ is a complex constant different from zero.

We first remark that an equation of the form

$$\text{(1.1)} \qquad z(x+1) = z(x) + a + \frac{b}{y(x)} ,$$

where a and b are complex constants and $b \neq 0$, is reduced to equation (E) by a simple transformation. In fact, equation (1.1) is transformed by

$$y(x) = z(x) / a$$

into equation (E) with $\lambda = b/a^2$ if $a \neq 0$, and by

$$y(x) = z(x)^2 / 2b$$

into equation (E) with $\lambda = 1/4$ if $a = 0$.

We next state our motivation of studying equation (E). The complex analytic theory of difference equations is not similar to that of differential equations in some points. One of the most different points is that the analytic difference equation does not have general existence theorems of analytic solutions analogous to those for the analytic differential equation . This fact compels one to prove the existence of solutions whenever a difference equation is given. Moreover analytic solutions of an analytic difference

This research was supported in part by NSF GP 27275.

equation are usually obtained under suitable asymptotic conditions, for example, under an asymptotic condition as x tends to ∞ along the real positive or negative axis. This situation necessarily gives to solutions to be obtained a local character. If we want to get global solutions, then we have to continue known local solutions analytically. However, the analytic continuation of local solution is not easy for nonlinear difference equations. In a previous paper [1], we obtained two kinds of local solutions of an equation

$$(1.2) \qquad y(x+1) = y(x) + 1 + \frac{\lambda}{y(x)} + \frac{\mu}{y(x)^2} + \frac{\nu}{y(x)^3} + \cdots ,$$

where the series $\lambda z^{-1} + \mu z^{-2} + \nu z^{-3} + \cdots$ converges for large values of z. In order to get global solutions of (1.2) by analytic continuation of the local solutions obtained, we must specify equation (1.2). For this purpose we take the simplest way of truncating the series $\lambda z^{-1} + \mu z^{-2} + \nu z^{-3} + \cdots$ to the first term λz^{-1} only. Then we get equation (E). Our motivation is to examine what phenomena happen for the global solutions of (E) which are obtained from local solutions by analytic continuation. This motivation gives to our study an experimental meaning in the global theory of an analytic difference equation.

We shall restrict ourselves to solutions meromorphic in $|x| < \infty$ and shall call such solutions meromorphic ones. The meromorphic solutions of (E) form, of course, a special class of solutions of (E).

2. The conformal map $w = z + 1 + \lambda z^{-1}$

Before studying equation (E), we shall make preliminary considerations about the formal map f defined by

$$f(z) = z + 1 + \lambda z^{-1}$$

for later use. The circle $|z| = \sqrt{|\lambda|}$ is mapped by f onto the straight segment L connecting the points $1 + 2\sqrt{\lambda}$ and $1 - 2\sqrt{\lambda}$. In particular,

$$f(\sqrt{\lambda}) = 1 + 2\sqrt{\lambda} \quad \text{and} \quad f(-\sqrt{\lambda}) = 1 - 2\sqrt{\lambda}.$$

The domains $0 \leqq |z| < \sqrt{|\lambda|}$ and $\sqrt{|\lambda|} < |z| \leqq \infty$ are both mapped one-to-one onto the domain $\overline{\mathbb{C}} - L$, where $\overline{\mathbb{C}}$ denotes the Riemann sphere. f has two fixed points $z = -\lambda, \infty$:

$$f(-\lambda) = -\lambda \quad \text{and} \quad f(\infty) = \infty.$$

Note that $f(-1) = -\lambda$ and $f(0) = \infty$ and that the equation $f(z) = -\lambda$ admits the only double root $z = -1$ if and only if $\lambda = 1$.

For n-th iterate f^n of f (f^0 = identity map and $f^1 = f$) and for any $\alpha \in \overline{\mathbb{C}}$ we define

$$A_n(\alpha) = \{\beta ; f^n(\alpha) = \beta\}$$

and then define

$$A(\alpha) = \bigcup_{r=0}^{\infty} A_n(\alpha).$$

It is clear that $A(\alpha)$ contains always the point α. The following proposition will play an important role in the next section.

Proposition 2.1. The set $A(\alpha)$ is reduced to the set $\{\alpha\}$ if and only if $\lambda = 1$ and $\alpha = -1$. If either $\lambda \neq 1$ or $\alpha \neq -1$, then $A(\alpha)$ contains at least three points.

Proof. The first assertion is evident.

To prove the second, we suppose that

$$\lambda \neq 1 \quad \text{or} \quad \alpha \neq -1. \tag{2.1}$$

Consider first the set

$$A_1(\alpha) = \{\beta ; f(\beta) = \alpha\}.$$

We have two possibilities:

Case 1. $A_1(\alpha)$ consists of two mutally distinct values β_1 and β_2.

Case 2. $A_1(\alpha)$ consists of a point β only.

We shall examine the first case. If neither β_1 nor β_2 is equal to α, then $A_1(\alpha)$, and hence $A(\alpha)$, contains at least three points α, β_1 and β_2. Assume that one of β_1 and β_2, say β_2, coincides with α. Then

$$A_1(\alpha) = \{\alpha, \beta_1\} . \tag{2.2}$$

We see that $A_1(\beta_1)$ does not contain α. In fact, if $\alpha \in A_1(\beta_1)$, then $f(\alpha) = \beta_1$. On the other hand, we have from (2.2)

$$f(\alpha) = \alpha .$$

It follows that $\alpha = \beta_1$, which contradicts. We next see that $A_1(\beta_1)$ contains a point different from β_1. In fact, if $A_1(\beta_1) = \{\beta_1\}$, then we have

$$\lambda = 1 \text{ and } \beta_1 = -1$$

and hence

$$\lambda = 1 \text{ and } \alpha = -1,$$

which contradicts (2.1). We thus conclude that $A_1(\beta_1)$ contains a point γ different from α and β_1. Since $A(\alpha) \supset A_1(\alpha) \cup A_1(\beta_1)$, $A(\alpha)$ contains the mutally distinct points α, β_1 and γ.

Case 2. We remark that $\alpha \neq \beta$. In fact, if $\alpha = \beta$, then we have $\lambda = 1$ and $\alpha = -1$. The condition $A_1(\alpha) = \{\beta\}$ means that the equation

$$z + 1 + \frac{\lambda}{z} = \alpha \quad \text{or} \quad z^2 + (1 - \alpha) z + \lambda = 0$$

has the only double root $z = \beta$. Therefore we have

$$\alpha = 1 \pm 2\sqrt{\lambda} \text{ and } \beta = \pm\sqrt{\lambda} .$$

Consider the set $A_1(\beta)$. We see that $\beta \notin A_1(\beta)$. In fact, if $\beta \in A_1(\beta)$, then $f(\beta) = \beta$. On the other hand, $f(\beta) = \alpha$. It follows that $\alpha = \beta$, which is a contradiction. The proof will be completed if we show that the case $A_1(\beta) = \{\alpha\}$ does not occur. In fact, $A_1(\beta) \neq \{\alpha\}$ means that $A_1(\beta)$ contains a point $\gamma \neq \alpha$. Since $\gamma \neq \beta$, $A(\alpha)$ contains the three points α, β and γ. Suppose that $A_1(\beta) = \{\alpha\}$. It follows from $f(\alpha) = \beta$ that

$$1 \pm 2\sqrt{\lambda} + 1 + \frac{\lambda}{1 \pm 2\sqrt{\lambda}} = \pm\sqrt{\lambda},$$

whence

$$3\lambda \pm 5\sqrt{\lambda} + 2 = 0. \tag{2.3}$$

It follows from the fact that α is a double root of $f(z) = \beta$ that

$$3\lambda \pm 2\sqrt{\lambda} - 1 = 0. \tag{2.4}$$

From (2.3) and (2.4) we have

$$\lambda = 1,$$

from which follows

$$\alpha = 3,\ \beta = 1 \quad \text{or} \quad \alpha = -1,\ \beta = -1.$$

The first case contradicts $f(\alpha) = \beta$, and the second case does (2.1).

The proposition is thus established.

We like to make the following definition. For an arbitrary value $\alpha \in \overline{\mathbb{C}}$, we say that a sequence $\{\alpha_n\}_{n=0}^{\infty}$ is an <u>α- sequence</u> if

$$\alpha_0 = \alpha \quad \text{and} \quad \alpha_{n-1} = f(\alpha_n) \ (n=1, 2, \ldots).$$

There is in general an infinite number of α-sequences for a given α.

3. Properties of meromorphic solutions

In this section we shall derive several properties of meromoprhic solutions of (E) assuming their existence.

From $f(\infty) = \infty$ and $f(-\lambda) = -\lambda$ we have the following propositions

Proposition 3.1. If a meromorphic solution $y(x)$ of (E) has a pole at $x = x_0$, then $y(x)$ has poles at

$$x = x_0 + 1, x_0 + 2, \dots .$$

Proposition 3.2. If a meromorphic solution $y(x)$ of (E) takes the value $-\lambda$ at $x = x_0$, then $y(x)$ takes $-\lambda$ at

$$x = x_0 + 1, x_0 + 2, \dots .$$

From $f(0) = \infty$ and $f(-1) = -\lambda$ we have the following.

Proposition 3.3. If a meromorphic solution $y(x)$ of (E) has a zero at $x = x_0$, then $y(x)$ has poles at

$$x = x_0 + 1, x_0 + 2, \dots .$$

Proposition 3.4. If a meromorphic solution $y(x)$ of (E) takes the value -1 at $x = x_0$, then $y(x)$ takes $-\lambda$ at

$$x = x_0 + 1, x_0 + 2, \dots .$$

It is clear that equation (E) admits the solution $y(x) \equiv -\lambda$, which we shall call the trivial solution.

Theorem 3.1. Equation (E) does not admit a nontrivial rational function solution. Therefore any nontrivial meromorphic solution is a transcendental function.

Proof. Suppose that $y(x)$ is a nontrivial rational function solution of (E). Then $y(x)$ has necessarily a pole or a zero at a point $x_0 \neq \infty$. It follows from Prop. 3.1 or 3.3 that $y(x)$ has an infinite number of poles at

$$x = x_0 + 1, x_0 + 2, \dots ,$$

which is a contradiction.

The following proposition is evident from the definition of $A(\alpha)$.

Proposition 3.5. If a meromorphic solution $y(x)$ of (E) does not take a value α, then $y(x)$ never takes any value in $A(\alpha)$.

Combining Prop. 2.1 and Prop. 3.5, we obtain the

Theorem 3.2. Any nontrivial meromorphic solution takes every value if $\lambda \neq -1$ and takes every value other than -1 if $\lambda = 1$.

Proof. Suppose that $\lambda \neq 1$ or $\alpha \neq -1$. Then $A(\alpha)$ contains at least three points. If therefore a nontrivial meromorphic solution $y(x)$ does not take the value α, then $y(x)$ does not take at least three points. Since $y(x)$ is necessarily a transcendental function, this fact contradicts the big Picard theorem. The theorem is thus proved.

From the definition of an α - sequence we have the

Proposition 3.6. If a meromorphic solution $y(x)$ of (E) takes a value α at x_0, then the sequence of values

$$\{y(x_0 - n)\}_{n=0}^{\infty}$$

is an α - sequence.

4. Existence of nontrivial meromorphic solutions

To prove the existence of nontrivial meromorphic solutions, we need the existence theorem of local solutions. Applying results obtained in the cited paper, we obtain the following propositions on local solutions.

Proposition 4.1. Equation (E) has a formal solution of the form

$$y(x) \sim x\ (1 + \sum_{j+k \geq 1} p_{jk}\, x^{-j} (x^{-1} \log x)^k),$$

where p_{10} is an arbitrary constant, $p_{01} = \lambda$ and other coefficients are determined as functions of p_{10} in a unique way.

We set $p_{10} = c$ and denote by $y(x, c)$ the corresponding formal solution.

To state existence theorem we shall use the following notation:

$D_\ell(R, \varepsilon)$ denotes the domain defined by

$$|x| > R, \ |\arg x - \pi| < \frac{\pi}{2} - \varepsilon \text{ or } \operatorname{Im}(xe^{-i\varepsilon}) > R \text{ or } \operatorname{Im}(x e^{i\varepsilon}) < -R$$

and $D_r(R, \varepsilon)$ denotes the domain defined by

$$|x| > R, \ |\arg x| < \frac{\pi}{2} - \varepsilon \text{ or } \operatorname{Im}(x e^{i\varepsilon}) > R \text{ or } \operatorname{Im}(xe^{-i\varepsilon}) < -R,$$

where R is a large positive number and ε is a small positive number.

Proposition 4.2. For an arbitrary c there is an actual solution $\varphi(x, c)$ (or $\Psi(x, c)$) of (E) with the following properties:

(i) $\varphi(x, c)$ (or $\Psi(x, c)$) is holomorphic in $D_\ell(R, \varepsilon)$ (or $D_r(R, \varepsilon)$),

(ii) $\varphi(x, c)$ (or $\Psi(x, c)$) is asymptotically developable into the formal solution $y(x, c)$ as x tends to ∞ through $D_\ell(R, \varepsilon)$ (or $D_r(R, \varepsilon)$).

Here ε can be taken arbitrarily but R depends on c and ε.

The following proposition is a uniqueness theorem.

Proposition 4.3. A solution $\varphi(x)$ of (E) which is holomorphic in $D_\ell(R, \varepsilon)$ (or $D_r(R, \varepsilon)$) and satisfies

$$\varphi(x) - x - \lambda \log x \to c \text{ as } x \to \infty, \ x \in D_r(R, \varepsilon) \text{ (or } D_\ell(R, \varepsilon)).$$

coincides with $\varphi(x, c)$ (or $\Psi(x, c)$).

It is clear that we can continue $\varphi(x, c)$ meromorphically into the whole complex plane $\mathbb{C}$ by making use of equation (E) itself. Thus we obtain from $\varphi(x,c)$ a meromorphic solution, which we denote by the same notation $\varphi(x, c)$.

A study of iteration of $f(z)$ gives us the following

Proposition 4.4. The ratio $\varphi(x, c)/x$ rests bounded in the domain D defined by

$$|x| > R, \ \operatorname{Re}(x) < 0 \text{ or } |\operatorname{Im}(x)| > R.$$

The growth of the meromorphic solution $\varphi(x, c)$ in D is very mild. Prop. 4.3 yields the

Proposition 4.5. For every c, we have

$$\varphi(x, c) = \varphi(x + c, 0).$$

5. Poles of $\varphi(x, c)$

The aim of this section is a study of behavior of $\varphi(x, c)$ in $\mathbb{C} - D$. For this purpose we shall focus our attention to poles of $\varphi(x, c)$.

Let $x = x_0$ be a pole of $\varphi(x, c)$. Then all the points

$$x_0 + 1,\ x_0 + 2,\ \ldots.$$

are poles of $\varphi(x, c)$ by Prop. 3.1, but not all

$$x_0 - 1,\ x_0 - 2, \ldots.$$

can be poles of $\varphi(x, c)$ by virtue of the holomorphy of $\varphi(x, c)$ in $D_r(R, \varepsilon)$. We can suppose that $\varphi(x, c)$ is holomorphic at $x_0 - 1,\ x_0 - 2, \ldots$. Consider the α - sequence

$$\{\varphi(x_0 - n, c)\}_{n=0}^{\infty} .$$

If n is sufficiently large, then $x_0 - n$ belongs to $D_r(R, \varepsilon)$. We see from the asymptotic expansion of $\varphi(x, c)$ that

$$\varphi(x_0 - n, c) \in D_r(R, \varepsilon) \quad \text{for } n \geqq n_0,$$

n_0 being a sufficiently large number. On the other hand, since x_0 is a pole of $\varphi(x, c)$, we can find a neighborhood U of x_0 such that the image of U under $\varphi(x, c)$ contains the domain $|x| > R$ and hence $D_r(R, \varepsilon)$. This implies that there is a sequence $\{x_0^{(n)}\}_{n=n_0}^{\infty}$, $x_0^{(n)} \in U$ such that

$$\varphi(x_0^{(n)}, c) = \varphi(x_0 - n, c) \qquad n \geqq n_0.$$

We can suppose without loss of generality that $x_0^{(n)}$ converge to x_0. Since

$$\varphi(x_0^{(n)} + n, c) = \varphi(x_0, c), \quad n \geqq n_0,$$

$x_0^{(n)} + n \ (n \geqq n_0)$ are all poles of $\varphi(x,c)$. From Prop. 3.1 we see that $\varphi(x,c)$ has an infinite number of sequences of poles:

$$x_0^{(n_0)} + n_0, \ x^{(n_0)} + n_0 + 1, \ldots.$$

$$x_0^{(n_0+1)} + n_0 + 1, \quad x_0^{(n_0+1)} + n_0 + 2, \ldots$$

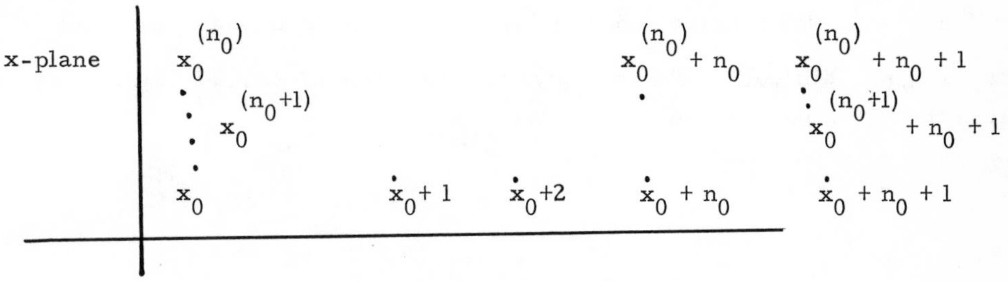

We may express this fact by saying that a sequence of poles of $\varphi(x,c)$ yields an infinite number of sequences of poles of $\varphi(x,c)$. Such a phenomenon does not occur for the linear difference equation.

6. Order of $\varphi(\cdot, c)$

We shall consider the order of the meromorphic function $\varphi(x,c)$ in the sense of Nevanlinna. By definition, if $T(r, \varphi(\cdot, c))$ denotes the characteristic function of $\varphi(x,c)$, then the order $\rho(\varphi(\cdot, c))$ of $\varphi(\cdot, c)$ is given by

$$\rho(\varphi(\cdot, c)) = \overline{\lim_{r \to \infty}} \frac{\log T(r, \varphi(\cdot, c))}{\log r} .$$

$T(r, \varphi(\cdot, c))$ is the sum of the proximity function $m(r, \varphi(\cdot, c))$ and the counting function $N(r, \varphi(\cdot, c))$:

$$T(r, \varphi(\cdot, c)) = m(r, \varphi(\cdot, c)) + N(r, \varphi(\cdot, c)),$$

where

$$m(r, \varphi(\cdot, c)) = \int_0^{2\pi} \log \sqrt{1+\varphi(r, e^{i\theta}, c)^2} \, d\theta \;,$$

$$N(r, \varphi(\cdot, c)) = \int_0^{r} \frac{n(t, \varphi(\cdot, c)) - n(0, \varphi(\cdot, c))}{t} \, dt$$

$$+ n(0, \varphi(\cdot, c)) \log r,$$

$n(t, \varphi(\cdot, c))$ being the number of poles of $\varphi(\cdot, c)$ lying in $|x| \leqq t$.

It seems that the behavior of $\varphi(x, c)$ in the domain $\mathbb{C} - D$ is very complicated in contrast to the behavior in D. This fact makes an estimation of $m(r, \varphi(\cdot, c))$, difficult. We shall give an estimation of $N(r, \varphi(\cdot, c))$, which derives the following theorem.

Theorem 6.1. We have

$$\rho(\varphi(\cdot, c)) \geqq 2.$$

Proof. By the result in the preceding section, $\varphi(x, c)$ has an infinite number of sequences of poles

$$
\begin{array}{l}
a_0,\ a_0 + 1,\ a_0 + 2, \ldots\ , \\
a_1 + 1,\ a_1 + 2,\ a_1 + 3, \ldots \\
a_2 + 2,\ a_2 + 3,\ a_2 + r, \ldots \\
\vdots \qquad \vdots \qquad \vdots
\end{array}
$$

where the sequence $a_1, a_2, \ldots$ converges to a_0. We may suppose without loss of generality that

$$|a_k - a_0| < 1/4 \quad \text{for } k = 1, 2, \ldots\ .$$

We see that if we have

$$|a_0 + k| + \frac{1}{4} \leqq t < |a_0 + k + 1| + \frac{1}{4},$$

then we have

$$n(t, \varphi(\cdot\ , c)) \geqq \sum_{m=1}^{k+1} m = \frac{1}{2}(k+1)(k+2),$$

from which we can deduce

$$n(t, \varphi(\cdot\ , c)) \geqq \begin{cases} 0 & (0 \leqq t < |a_0| + \frac{1}{4}) \\ \frac{1}{2}(t - |a_0| - \frac{1}{4})(t - |a_0| + \frac{3}{4}) & (|a_0| + \frac{1}{4} \leqq t). \end{cases}$$

Simple calculations show that we have

$$N(r, \varphi(\cdot\ , c)) \geqq \frac{1}{4} r^2 + c_1 r + c_2 \log r + c_3$$

for $r \geqq 0$, where c_1, c_2 and c_3 are constants depending on c and λ only. We can conclude from this estimation that

$$\rho(\varphi(\cdot\ , c)) \geqq 2.$$

This proves the theorem.

7. Open problems

We shall list several open problems concerning equation (E).

(I) Is the order $\rho(\varphi(\cdot\ , c))$ of $\varphi(x, c)$ equal to two ?

(II) A transcendental function which does not satisfy any algebraic differential equation

$$F(x, y, y', \ldots, y^{(n)}) = 0$$

is called a hypertranscendental function over the differential field $\mathbb{C}(x)$ of rational functions, where F is a polynomial in all arguments. It is well-known that the Γ- function $\Gamma(x)$ is hypertranscendental over $\mathbb{C}(x)$.

Is the function $\varphi(x, c)$ hypertranscendental over $\mathbb{C}(x)$?

(III) Let $y(x)$ be an arbitrary solution of (E). Then for any periodic function $\pi(x)$ with period 1, $y(x + \pi(x))$ is also a solution of (E). In fact, this follows from

$$y(x+1+\pi(x+1)) = y(x+\pi(x)+1)$$

$$= y(x+\pi(x)) + 1 + \frac{\lambda}{y(x+\pi(x))} .$$

If $\pi(x)$ is an entire periodic function with period 1, then $\varphi(x+\pi(x), c)$ is a meromorphic solution of (E). We can proposed the following problem.

To find meromorphic solutions which are not expressible as $\varphi(x+\pi(x), c)$.

(IV) The solution $\Psi(x, c)$ stated in Prop. 4.2 is holomorphic in the domain $D_r(R, \varepsilon)$ and is continuable into the whole plane $\mathbb{C}$.

Is the solution $\Psi(x, c)$ always a multiple-valued solution ?

References

1. T. Kimura, "On the iteration of analytic functions," Funkcialaj Ekvacioj, 14(1971), 197-238.

Branching of Periodic Solutions

W. S. Loud

1. Introduction

Consider a second-order real ordinary differential equation

$$x'' + g(x) = Ef(t) \tag{1.1}$$

where $g(x)$ is a sufficiently regular odd function with $xg(x) > 0$ for $x \neq 0$, and $f(t)$ is a continuous (or perhaps piecewise continuous) 2π-periodic function. We also assume that $f(t)$ is an even function which is odd-harmonic (i.e. $f(t + \pi) \equiv - f(t)$).

Such equations as (1.1) occur in nonlinear mechanics, and it is of interest to learn about their periodic solutions. The very existence of such solutions is a substantial problem, since there are simple linear equations of the form (1.1) which have no periodic solutions. Other questions which arise after the establishment of existence are those of stability and constructibility.

When equation (1.1) is linear, where $g(x) = k^2x$, there is a unique 2π - periodic solution provided that k^2 is not the square of an integer. What is more, this unique periodic solution is an even function which is odd-harmonic. This leads to the expectation that periodic solutions of (1.1) in general will be even and odd-harmonic, and in certain cases, this proves to be the case. However, other phenomena are known to occur with nonlinear equations. For some values of E, (1.1) may have more than one 2π - periodic solution, with the number of such solutions changing abruptly at some values of E.

The research for this paper was supported in part by the U.S. Army Research Office (Durham) Grant No. DA-ARO(D)-31-124-G1154.

Another phenomenon is the appearance for some values of E of periodic solutions which do not have the properties of being both even and odd-harmonic. Such phenomena are well known to engineers, but their knowledge of them is often based on analogue computer studies. It is the purpose of this discussion to give a rigorous mathematical investigation of such phenomena.

One reason that the topic of branching of periodic solutions is appropriate at a symposium in honor of H. L. Turrittin is that one of the first mathematical associations of the author with Professor Turrittin was in connection with the mathematical investigation of such a phenomenon.

The equation in question was

$$x'' + ax' + x + \beta x^3 = \sin t \tag{1.2}$$

where the parameter a is fixed as a small positive quantity. Analogue computer studies had suggested that for small positive values of β there was a single odd-harmonic 2π - periodic solution of (1.2). This is qualitatively the same situation as occurs in the linear situation with $\beta = 0$. However, as the parameter β was increased through positive values, the stable periodic solution was noticed to be no longer odd-harmonic. Its graph had roughly the shape of $x = \cos t + \varepsilon \cos 2t$. cf. [8].

Professor Turrittin and his student W. J. A. Culmer were able [7] to construct an example of a differential equation related to (1.2) which could be shown to possess a nonsymmetric periodic solution of the type desired. Later [1] C.A. Harvey constructed further examples of nonsymmetric solutions of equations of the type (1.1).

The following sections are a further elaboration of the branching phenomena described above with an attempt at explaining the cause of the presence or absence of such behavior in terms of the behavior of the function $g(x)$.

2. A Family of Periodic Solutions

The first step in the investigation of branching of various types is the establishment of the existence of even, odd-harmonic, 2π - periodic solutions of (1.1) for many values of the parameter E. In particular it is of interest to know of such solutions in a situation that is not a perturbation of a linear problem or of an autonomous problem. Such nonperturbational problems present considerable difficulty, and it is only by exploiting the symmetry properties of the equation that progress was made.

Consider first a situation in which $g(x)$ is piecewise linear:

$$g(x) = \begin{cases} k_2^2 x + (k_1^2 - k_2^2), & x \geqq 1 \\ k_1^2 x, & |x| \leqq 1 \\ k_2^2 x + (k_2^2 - k_1^2), & x \leqq -1 \end{cases}$$

In this problem, let $f(t) = \cos t$. It is possible to find solutions of (1.1) for this case with $x'(0) = 0$, $x(\pi/2) = 0$ for some values of E. If we assume that for this solution $x(t) \geqq 1$ for $0 \leqq t \leqq t_1$ and $0 \leqq x(t) \leqq 1$ for $t_1 \leqq t \leqq \pi/2$, we can find, using t_1 as parameter, expressions for the parameter E, for $A = x(0)$, and $V = x'(\pi/2)$. For suitable restrictions on k_1 and k_2 such a solution will exist for an interval of values of t_1, and this will often correspond to one or more intervals of values of E. The symmetry of the equation guarantees that any solution having $x'(0) = x(\pi/2) = 0$ will be an even odd-harmonic, 2π - periodic solution. cf. [4].

Suppose that there is a 2π - periodic solution $x(t, E)$, for some value of E. Standard perturbation theory says that if the variation equation

$$y'' + g'(x(t, E))\, y = 0 \qquad (2.1)$$

has no nontrivial 2π - periodic solution, there will exist for nearby values of E a unique 2π - periodic solution. For the piecewise linear system, the coefficient $g'(x(t, E))$ is piecewise constant, π-periodic, and even. It changes value just once in the interval $0 \leq t \leq \pi/2$, namely at $t = t_1$. It can be shown that (2.1) will have no nontrivial 2π - periodic solution if there are no square integers between k_1^2 and k_2^2. In this case, the variation equation will never have a nontrivial 2π-periodic solution, so that the family $x(t, E)$ can be extended to all values of E.

This same result [3] was extended to more general $g(x)$ than piecewise linear. Suppose that along with the assumptions stated earlier $g'(x)$ is positive, continuous, and bounded away from square integers. Then equation(1.1) has a unique 2π-periodic solution for all values of E, and this solution is odd-harmonic and even. It is possible to relax the hypothesis to the requirement that the range of $g'(x)$ be bounded away from odd square integers. In this case there is a unique even, odd-harmonic, 2π-periodic solution, but there may exist other 2π-periodic solutions.

The proof of these assertions is given in [3]. The idea of the proof is based on continuity considerations. If a solution of (1.1) has $x'(0) = 0$ and $x(\pi/2) = 0$, the symmetry properties of the equation then imply that the solution is 2π-periodic, even, and odd-harmonic. Let $x(t, A)$ be that solution of (1.1) which has initial conditions $x(0) = A$, $x'(0) = 0$, and let $F(A) = x(\pi/2, A)$. If $g'(x)$ is bounded away from odd integers,the function $F(A)$ is strictly monotonic and varies from $-\infty$ to $+\infty$. Hence there is exactly one value of A for which $F(A) = 0$, so that there is exactly one even, odd-harmonic, 2π-periodic solution. If $g'(x)$ is bounded away from all square integers, it follows that any 2π - periodic solution of (1.1) must have $x'(0) = x'(\pi) = 0$. If $G(A)$ is defined to be $x'(\pi, A)$, it will follow that $G(A)$ is strictly monotonic, so that there is at most one value of A with $G(A) = 0$, and hence at most one 2π-periodic solution of (1.1). But from the earlier argument we know there is a

2π-periodic, even, odd-harmonic solution which must be the unique periodic solution in this case.

The above results were generalized by D. E. Leach [2]. Leach assumes only that $g'(x)$ is continuous, positive, and bounded away from square integers, and that $f(t)$ is continuous and 2π-periodic. It is not assumed that $g(x)$ is odd, and it is not assumed that $f(t)$ is even or odd-harmonic. The result is that there is still a unique 2π-periodic solution of (1.1).

Informally stated, the results of this section show that unless the range of $g'(x)$ contains a square integer, the behavior of (1.1) so far as periodic solutions are concerned is qualitatively the same as for linear equations. This might be described by saying that for this case the nonlinearity is not very great.

3. Harmonic Branching

In this section we consider the effect of removing the essential hypothesis of the preceding section. When the range of $g'(x)$ is bounded away from square integers, the variation equation (2.1) will never have a nontrivial 2π-periodic solution. Branching phenomena occur exactly when this equation does have a nontrivial 2π-periodic solution. Of course the presence of square integers in the range of $g'(x)$ does not guarantee the presence of such nontrivial periodic solutions, so the hypotheses for branching will be the existence of nontrivial periodic solutions of the variation equation.

If $x(t)$ is an even, odd-harmonic, 2π-periodic solution of (1.1), the coefficient $g'(x(t))$ in the variation equation

$$y'' + g'(x(t))\, y = 0 \tag{3.1}$$

is a π-periodic even function. Nontrivial periodic solutions of (3.1) fall into four cases:

I 2π-periodic, odd, and odd-harmonic,

II 2π-periodic, even, and odd-harmonic,

III π-periodic and even,

IV π-periodic and odd.

Cases I and II can occur together as can cases III and IV. This is the phenomenon of co-existence, and implies that all solutions of (3.1) are periodic. We rule this out in our considerations. In each case where (3.1) has a nontrivial 2π-periodic solution, we assume there is just one solution (along with its constant multiples) which is periodic.

In the piecewise linear case mentioned above cases I and II correspond to odd square integers lying between k_1^2 and k_2^2, while cases III and IV correspond to even square integers lying between k_1^2 and k_2^2.

The four cases of periodic solutions of (3.1) in the piecewise linear case result in four qualitatively different types of branching of the family $x(t, E)$. The only value of E which gives rise to a type I periodic solution is $E = 0$. The solution $x(t, 0)$ is of course a 2π-periodic, even, odd-harmonic solution of the autonomous equation $x'' + g(x) = 0$. Consequently in the neighborhood of this solution are members of the family $x(t, E)$ for small nonzero E and also nearby translations in time of the solution $x(t, 0)$ for E held fixed at zero. I.e. the branching of solutions from the family $x(t, E)$ is the translations $x(t+s, 0)$ of the solution $x(t, 0)$.

A type II periodic solution of (3.1) corresponds to an endpoint of an interval of values of E for which periodic solutions $x(t, E)$ exist. Recall that the transition time t_1 was used as a parameter. The parameter E as a function of t_1 has an extremum corresponding to a type II periodic solution, so that for values of E on one side of the endpoint there are two members of the family $x(t, E)$, corresponding to values of t_1 just above and just below the critical value. Thinking of E as the parameter, as E passes through the critical value, there suddenly appear two 2π-periodic, even, odd-harmonic solutions (or if the passing is in the reverse direction, two different periodic solutions come together and disappear). This is the well-known jump phenomenon which has been known for a long time about Duffing's equation.

A type III periodic solution of (3.1) corresponds to a branching of the type described in the introduction. For E on one side of the critical value, there suddenly appear two new solutions, which are still even and 2π-periodic but are not odd-harmonic. (The former result that when $g'(x)$ is bounded away from odd square integers, there is a unique odd-harmonic solution applies here.) The nonsymmetric solutions have graphs that are suggested by the graph of $x = \cos t + \varepsilon \cos 2t$.

A type IV periodic solution of (3.1) corresponds to a similar branching to the type III case. The new solutions are not even functions, but are still odd functions about $t = \pi/2$. Their graphs are suggested by the graph of $x = \cos t + \varepsilon \sin 2t$.

A more detailed discussion of the piecewise linear case, with proofs, is given in [4].

Similar results can be obtained for more general $g(x)$. Cases in which branching of types III and IV can be proved are given in [5]. Here it is necessary to assume that $g'(x)$ is confined between two consecutive odd square integers and that $g'(x)$ is monotonic increasing for positive x and such that the intermediate even square integer is definitely contained in its range. We know then that the family $x(t, E)$ of 2π-periodic, even, odd-harmonic solutions exists, and it can be shown that for some value of E the variation equation has a type III nontrivial periodic solution and for some other value of E (we exclude co-existence) there is a type IV nontrivial periodic solution.

The following is an indication of the proof for the type III case. For certain details of the computations see [5]. Let E_o be a value of E for which the variation equation has a type III periodic solution, but not a type IV periodic solution. We shall show that at $E = E_o$ the family $x(t, E)$ has a branching of the type described for the piecewise linear case.

Let $x(t, \xi, \eta, \varepsilon)$ be that solution of (1.1) with $E = E_o + \varepsilon$ and initial conditions $x(0, \xi, \eta, \varepsilon) = x(0, E_o) + \xi$, $x'(0, \xi, \eta, \varepsilon) = \eta$. Then let $F(\xi, \eta, \varepsilon) = x(2\pi, \xi, \eta, \varepsilon) - x(0, E_o) - \xi$, and $G(\xi, \eta, \varepsilon) = x'(2\pi, \xi, \eta, \varepsilon) - \eta$. The solution $x(t, \xi, \eta, \varepsilon)$ is 2π-periodic if and only if $F(\xi, \eta, \varepsilon) = G(\xi, \eta, \varepsilon) = 0$. Since $x(t, 0, 0, 0) = x(t, E_o)$, which is 2π-periodic, $F(0, 0, 0) = G(0, 0, 0) = 0$. We shall show how the solutions of the system $F = G = 0$ for ξ and η as functions of ε behave for ε near zero.

The partial derivatives of F and G at $(0, 0, 0)$ are computed by investigating the partial derivatives of $x(t, \xi, \eta, \varepsilon)$ and $x'(t, \xi, \eta, \varepsilon)$ at $t = 2\pi$. The computations for higher derivatives are complicated, and we give only the needed results. It is found in the present case that at $(0, 0, 0)$: the partial derivatives F_ξ, F_ε, $F_{\xi\xi}$, G_ξ, G_η, G_ε, $G_{\xi\xi}$, $G_{\xi\eta}$ are all zero; $F_\eta \neq 0$; $G_{\xi\varepsilon}$, $G_{\varepsilon\varepsilon}$, $G_{\xi\xi\xi}$, and the combination $F_{\xi\varepsilon} + G_{\eta\varepsilon}$ are in general nonzero, and we assume that they are all indeed nonzero. The partial derivatives not mentioned in the above listing are not critical in determining existence of the branching solutions.

Because $F_\eta \neq 0$, the equation $F(\xi, \eta, \varepsilon) = 0$ may be solved near $\xi = \varepsilon = 0$ for η as a function of ξ and ε. Let the result be $\eta = H(\xi, \varepsilon)$. The above results give that at $(0, 0)$ $H = H_\xi = H_\varepsilon = 0$. Now define $J(\xi, \varepsilon)$ by the formula

$$J(\xi, \varepsilon) = G(\xi, H(\xi, \varepsilon), \varepsilon).$$

The above results give that at $(0, 0)$ $J = J_\xi = J_\varepsilon = J_{\xi\xi} = 0$, $J_{\xi\varepsilon}$, $J_{\varepsilon\varepsilon}$, and $J_{\xi\xi\xi}$ are nonzero. Hence the initial nonvanishing terms of the Taylor series for J are

$$\frac{1}{6} J_{\xi\xi\xi} \xi^3 + J_{\xi\varepsilon} \xi\varepsilon + \frac{1}{2} J_{\varepsilon\varepsilon} \varepsilon^2.$$

It is these terms which determine the asymptotic behavior for ε near zero of the solutions of $J(\xi, \varepsilon) = 0$ for ξ as a function of ε. They imply that near $(0, 0)$ the locus of the equation $J(\xi, \varepsilon) = 0$ consists of two curves, one tangent to the line $2J_{\xi\varepsilon}\xi + J_{\varepsilon\varepsilon}\varepsilon = 0$, the other tangent to the line $\varepsilon = 0$ and approximated initially by the parabola $J_{\xi\xi\xi}\xi^2 + 6J_{\xi\varepsilon}\varepsilon = 0$.

Hence the solutions of $F(\xi, \eta, \varepsilon) = G(\xi, \eta, \varepsilon) = 0$ have the expressions for small ε:

$$\xi = -\frac{J_{\varepsilon\varepsilon}}{2J_{\xi\varepsilon}}\varepsilon + o(\varepsilon), \qquad \eta = o(\varepsilon);$$

$$\xi = \pm\sqrt{\frac{-6J_{\xi\varepsilon}}{J_{\xi\xi\xi}}}\,\varepsilon^{\frac{1}{2}} + o(\varepsilon^{\frac{1}{2}}), \quad \eta = o(\varepsilon^{\frac{1}{2}}),$$

or

$$\xi = \pm\sqrt{\frac{6J_{\xi\varepsilon}}{J_{\xi\xi\xi}}}\,(-\varepsilon)^{\frac{1}{2}} + o((-\varepsilon)^{\frac{1}{2}}), \quad \eta = o((-\varepsilon)^{\frac{1}{2}}).$$

Note that according as the sign of $J_{\xi\varepsilon}J_{\xi\xi\xi}$ is negative or positive, the second solution exists for small positive ε only or for small negative ε only.

For each nonzero ε there exist, therefore, either three solutions or one solution of the system $F = G = 0$. When three exist, they are distinct. For each ε the ξ and η give initial conditions for 2π-periodic solutions of (1.1) with $E = E_o + \varepsilon$. Now since $x(t, E)$ is such a solution for all values of E, its initial conditions must correspond to the one solution, if there is only one, and to one of the three solutions, if there are three. Calculation will show that the initial conditions for $x(t, E)$, correspond to the first solution, in which $\xi = o(\varepsilon)$. The earlier reasoning gave $\eta = o(\varepsilon)$, but since $x(t, E)$ is an even function, $\eta \equiv 0$.

Let us now consider the two new solutions which come into existence at $E = E_o$ and exist for E on one side of E_o. Neither of them is $x(t, E)$, and hence neither of them is both even and odd-harmonic, since there is known to be only one solution with these properties, and $x(t, E)$ is that one. Let $x_1(t)$ be one of the two new solutions. Symmetry properties of the equation imply that $x_2(t) = x_1(-t)$, $x_3(t) = -x_1(t+\pi)$, and $x_4(t) = -x_1(\pi-t)$ are also solutions. None of these four can be even and odd-harmonic for then all four would have to be the same as $x(t, E)$. On the other hand they can not all four be distinct, since there are only two periodic solutions which are 2π - periodic but not both even and odd-harmonic. The four solutions must consist of two identical pairs. Since at $t = 0$, $x_1 = x_2$ and $x_3 = x_4$, it must be that $x_1 \equiv x_2$ and $x_3 \equiv x_4$. This accounts for the two new solutions and incidentally proves they are both even functions. The expansions in powers of ε for the new solutions are

$$(3.2) \qquad x = x(t, E_0) \pm \sqrt{\left|\frac{6J_{\xi\varepsilon}}{J_{\xi\xi\xi}}\right|} \; |\varepsilon|^{\frac{1}{2}} \varphi(t) + o(|\varepsilon|^{\frac{1}{2}})$$

it being understood that ε is taken only on the appropriate side of zero. In (3.2) $\varphi(t)$ is the π-periodic even solution of the variation equation

$$y'' + g'(x(t, E_0))\, y = 0,$$

with $\varphi(0) = 1$.

It does not seem possible to say anything truly definite about the stability of the solutions of types III and IV just discussed. All that can be said is that the solutions $x(t, E)$ are stable on one side of E_0 and unstable on the other, while the new solutions have opposite stability to $x(t, E)$ on whichever side of E_0 they happen to exist. In this statement stability of the solution is to mean that the corresponding solution of the damped equation

$$x'' + cx' + g(x) = Ef(t)$$

where c is positive and sufficiently small, is asymptotically stable.

4. Subharmonic Branching

It is also possible to have other periodic solutions branching from the family $x(t, E)$. Suppose that for some value $E = E_o$ the characteristic multipliers of the variation equation

$$y'' + g'(x(t, E_o))\ y = 0 \tag{4.1}$$

are $\exp\ \{\pm i\ p\ \pi/q\}$ where p and q are relatively prime positive integers with $0 < p < q$. The coefficient $g'(x(t, E_o))$ is still even and π-periodic. The solutions of (4.1) are all $2q\pi$-periodic. It seems reasonable to expect that there will be solutions of period $2q\pi$ near to $x(t, E_o)$ for values of E near to E_o. A solution whose period is a multiple of the fundamental period of $f(t)$ is called a subharmonic.

A subharmonic solution to (1.1) inevitably involves a number of related subharmonic solutions. In fact, if $x(t)$ is any subharmonic of least period $2q\pi$, there are always $2q-1$ other solutions: $x(t+2\pi), x(t+4\pi), \ldots, x(t+(2q-2)\pi)$, $-x(t+3\pi), \ldots, -x(t+(2q-1)\pi)$. These are all distinct if either of p and q is even. If p and q are both odd, there are only q distinct solutions in the family, since $-x(t+q\pi) \equiv x(t)$.

Existence of such solutions can be shown by methods similar to those mentioned above. This time $F(\xi, \eta, \varepsilon) = x(2q\pi, \xi, \eta, \varepsilon) - x(0, E_o) - \xi$, and $G(\xi, \eta, \varepsilon) = x'(2q\pi, \xi, \eta, \varepsilon) - \eta$. The solution $x(t, \xi, \eta, \varepsilon)$ is a subharmonic of period $2q\pi$ if and only $F = G = 0$. Again $F(0, 0, 0) = G(0, 0, 0) = 0$ since $x(t, E_o)$ has period 2π and hence period $2q\pi$.

The computations, particularly for all but the lowest values of q are very extensive. The author has carried out the computations completely only for the two cases $p/q = 1/3$ and $p/q = 1/2$. It seems that it is necessary to compute partial derivatives of F and G through order $q-1$ if p and q are both odd, and through order $2q-1$ if either of p or q is even. Because of the increasing complexity of the computation of partial derivatives, this does not seem to be the most desirable attack on the problem.

When $p/q = 1/3$, it is found that there are four 6π-periodic solutions both for $E < E_o$ and for $E > E_o$. One of these is, of course, $x(t, E)$ for E near E_o. The other three are a subharmonic $x_1(t)$ and its two translations $x_1(t+2\pi)$ and $x_1(t+4\pi)$. As was mentioned before, $x_1(t)$ is identical with $-x_1(t+3\pi)$. It is also possible to determine the stability of these solutions. The harmonics are stable; the subharmonics always unstable.

When $p/q = 1/2$, there are found to be, in addition to $x(t, E)$, a number of 4π-periodic solutions. There are two quite different possibilities. One possibility is that four exist for $E < E_o$ and four exist for $E > E_o$. For each side of E_o, if $x_1(t)$ is one solution, the others are $x_1(t+2\pi)$, $-x_1(t+\pi)$, and $-x_1(t+3\pi)$. All subharmonics in this situation are unstable. The second possibility is that for E on only one side of E_o there are eight 4π-periodic solutions. These are two solutions $x_1(t)$ and $x_2(t)$ together with the translations $x_1(t+2\pi)$, $-x_1(t+\pi)$, $-x_1(t+3\pi)$, $x_2(t+2\pi)$, $-x_2(t+\pi)$, and $-x_2(t+3\pi)$. In this situation one of x_1 and x_2 together with its three translations is stable and the other is unstable.

For the details of the proofs of these results, which are quite lengthy, the reader is referred to [6].

For the higher order cases, it appears that subharmonics will be half stable and half unstable.

References

1. C. A. Harvey, "Periodic Solutions of the Differential Equation $x''+g(x)=p(t)$," Contributions to Differential Equations, 1 (1962) 425-451.

2. D. E. Leach, "On Poincasé's Perturbation Theorem and a Theorem of W. S. Loud," J. Differential Equations, 7 (1970) 34-53.

3. W.S. Loud, "Periodic Solutions of Nonlinear Differential Equations of Duffing Type," Proceedings U.S. - Japan Seminar, Benjamin, New York (1967) 199-224.

4. W.S. Loud, "Branching Phenomena for Periodic Solutions of Nonautonomous Piecewise Linear Systems," Int. J. of Nonlinear Mechanics, 3 (1968) 273-293.

5. W.S. Loud, "Nonsymmetric Period Solutions of Certain Second Order Nonlinear Differential Equations", J. Differential Equations, 7(1970) 352-368.

6. W.S. Loud, "Subharmonic Solutions of Second Order Equations Arising near Harmonic Solutions", J. Differential Equations, 11(1972) 628-660.

7. H.L. Turrittin and W.J.A. Culmer, "A Peculiar Periodic Solution of a Modified Duffing's Equation," Ann. di Mat. Pura ed Appl. (IV) 44 (1957) 23-34.

8. D.L. Markusen, Dana Young, H. L. Turrittin, W.S. Loud, P.N. Hess W.J.A. Culmer, and E.J. Putzer, Minneapolis-Honeywell Aero Report AD5042-TR4, Vol. III "Analysis of Systems with Nonlinear Restoring Forces and Limiting,"1963.

Effective Solutions for Meromorphic Second Order Differential Equations

W. Jurkat, D. Lutz, and A. Peyerimhoff

In this note, we summarize a part of our results appearing in [5] by illustrating how these results may be applied to effectively solve differential equations of the form

(1) $$y'' + a(z)y' + b(z)y = 0,$$

when $a(z) = \sum_{i=0}^{\infty} a_i z^{-i}$, $b(z) = \sum_{i=0}^{\infty} b_i z^{-i}$, $a_0^2 \neq 4b_0$,

and both power series converge for $|z| > R$. The point at ∞ is an irregular singular point of (1) (see [3]; p. 111). It is convenient to write (1) in the system form

(2) $$Y' = \begin{bmatrix} 0 & 1 \\ -b(z) & -a(z) \end{bmatrix} Y,$$

by letting $y_1 = y$, $y_2 = y'$. The coefficient matrix of (2) is analytic at ∞ and the condition $a_0^2 \neq 4b_0$ means that the leading coefficient matrix has distinct eigenvalues. They are solutions of $\lambda^2 + a_0\lambda + b_0 = 0$ and we label them as $\lambda_1 \neq \lambda_2$. Next make the constant transformation $Y = \begin{bmatrix} 1 & 1 \\ \lambda_1 & \lambda_2 \end{bmatrix} X$ to obtain

(3) $$X' = A(z)X, \text{ where } A(z) = \begin{bmatrix} \lambda_1 & 0 \\ 0 & \lambda_2 \end{bmatrix} + \sum_{i=1}^{\infty} A_i z^{-i}$$

This work was supported in part by grants GP-19653 and GP-28149 from the National Science Foundation.

and the power series converges for $|z| > R$.

It is easy to compute a formal fundamental solution matrix for (3). There exists one of the form

$$\hat{X}(z) = F(z)z^{\Lambda'} \exp(\Lambda z), \tag{4}$$

where $\Lambda = \operatorname{diag}\{\lambda_1, \lambda_2\}$, $\Lambda' = \operatorname{diag} A_1 = \operatorname{diag}\{\lambda_1', \lambda_2'\}$, $F(z) = \sum_{n=0}^{\infty} F_n z^{-n}$, $F_0 = I$, and for $n \geq 1$, the F_n are uniquely calculated (recursively) in a well-known manner ([3]; pp. 141-147) from the equations

$$(n-1)F_{n-1} = F_n \Lambda + F_{n-1}\Lambda' - \sum_{i=0}^{n} A_i F_{n-i} (n \geq 1). \tag{5}$$

The complex numbers λ_1', λ_2' are determined by

$$\lambda_1' = (\lambda_2 - \lambda_1)^{-1}(b_1 + a_1\lambda_1) \text{ and } \lambda_2' = (\lambda_1 - \lambda_2)^{-1}(b_1 + a_1\lambda_2).$$

The actual solution of (3) (and therefore (1)) will come about by first transforming (3) by means of $X = T(z)W$ to a simplified form

$$W' = \left(\begin{bmatrix} \lambda_1 & 0 \\ 0 & \lambda_2 \end{bmatrix} + \frac{1}{z} \begin{bmatrix} \mu_1' & c \\ c' & \mu_2' \end{bmatrix} \right) W = B(z)\, W \tag{6}$$

and then solving (6) explicitly in terms of well-known functions. Of course, it is easy to see that by means of $X = F(z)U$ ($F(z)$ as in (4)), (3) is transformed into $U' = (\Lambda + \Lambda' z^{-1})\, U$, which has the form (6), however, the matrix $F(z)$ is generally just a formal series. What we shall do is select the parameters μ_1', μ_2', c, c' in (6) such that there exists an actual (matrix) function $T(z)$ which transforms (3) into (6) and then compute $T(z)$ effectively. Most of the time it is possible to use transformations of the form $T(z) = I + \sum_{1}^{\infty} T_i z^{-i}$, where the power series converges for $|z|$ sufficiently large. In this case $\mu'_1 = \lambda'_1$ and $\mu'_2 = \lambda'_2$. Sometimes, however, we will use transformations $T(z)$ which are meromorphic at ∞ and such that $\det T(z) \not\equiv 0$; in this case $\mu'_1 = \lambda'_1 + k_1$, $\mu'_2 = \lambda'_2 + k_2$, where k_1 and k_2 are intergers which need to be determined.

To proceed with the calculation of (6), we first make use of $F(z)$ to compute some additional quantities which are helpful in the selection of the parameters in (6). From $\lambda_1, \lambda_2, \lambda'_1, \lambda'_2$, we construct the diagonal matrices

$$K_n = \operatorname{diag}\{(-1)^n n^{\lambda'_2 - \lambda'_1}, n^{\lambda'_1 - \lambda'_2}\} (\lambda_2 - \lambda_1)^{-n} \Gamma(n), \; n \geq 1.$$

It turns out that $F_n K_n^{-1}$ has an asymptotic expansion in a power series in n^{-1} as $n \to \infty$, and in particular,

$$(7) \qquad \lim_{n \to \infty} F_n K_n = C \text{ exists and has the form } C = \begin{bmatrix} 0 & \gamma \\ \gamma' & 0 \end{bmatrix}.$$

Furthermore, $\gamma = 0$ iff the second column of $F(z)$ converges, while $\gamma' = 0$ iff the first column of $F(z)$ converges (for $|z|$ sufficiently large). Using γ, γ' computed from (7), we let μ denote the general solution of

$$(8) \qquad \cos 2\pi\mu = \cos \pi(\lambda'_2 - \lambda'_1) - 2\pi^2 \gamma\gamma'.$$

We consider the complex cosine function in (8) in order to have a solution μ always defined. It is sometimes convenient to have two additional parameters α_*, β_* available, which are defined by

$$(9) \qquad \alpha_* = \frac{1}{2}(\lambda'_2 - \lambda'_1) + \mu \text{ and } \beta_* = \frac{1}{2}(\lambda'_2 - \lambda'_1) - \mu.$$

We now "match" the given system (3) with a special system (6) by naming the parameters μ'_1, μ'_2, c, c'. Four cases are considered depending upon the zero, non-zero structure of the γ's in (7), i.e., depending upon the convergence, divergence property of the columns of $F(z)$:

Case (1) $(\gamma\gamma' \neq 0)$.

In this case, for all choices of μ satisfying (8), neither α_* nor β_* is an integer and we take (for any choice of μ)

(10) $\mu'_1 = \lambda'_1, \mu'_2 = \lambda'_2, \ c = \gamma\Gamma(1-\alpha_*)\Gamma(1-\beta_*), \ c' = \gamma'\Gamma(1+\alpha_*)\Gamma(1+\beta_*)$.

Case (2) $(\gamma \neq 0, \ \gamma' = 0)$.

(a). If $\lambda'_2 - \lambda'_1 \neq$ positive integer, then we take

(11) $\mu'_1 = \lambda'_1, \mu'_2 = \lambda'_2, \ c = \gamma\Gamma(1-\lambda'_2+\lambda'_1), \ c' = 0.$

(b). If $\lambda'_2 - \lambda'_1 = k$ (a positive integer), then we take

(12) $\mu'_1 = \lambda'_1, \mu'_2 = \lambda'_2 - k = \lambda'_1, \ c = \gamma, \ c' = 0.$

Case (3) $(\gamma = 0, \ \gamma' \neq 0)$.

(a). If $\lambda'_1 - \lambda'_2 \neq$ positive integer, then we take

(13) $\mu'_1 = \lambda'_1, \mu'_2 = \lambda'_2, \ c = 0, \ c' = \gamma'\Gamma(1+\lambda'_2-\lambda'_1).$

(b). If $\lambda'_1 - \lambda'_2 = k$ (a positive integer), then we take

(14) $\mu'_1 = \lambda'_1, \mu'_2 = \lambda'_2 + k = \lambda'_1, \ c = 0, \ c' = \gamma'.$

Case (4) $(\gamma = \gamma' = 0)$.

We select

(15) $\mu'_1 = \lambda'_1, \mu'_2 = \lambda'_2, \ c = c' = 0.$

A transformation $T(z)$ which takes (3) into (6) can be calculated as follows: Let $F_A(z) = F(z)$ denote the formal series in (4) for the system (3).

Let $F_B(z)$ denote the formal series in the formal fundamental solution $F_B(z) \operatorname{diag}\{z^{\mu'_1}, z^{\mu'_2}\} \exp(\Lambda z)$ for (6), where $F_B(z)$ has leading term I. In cases (1), (2a), (3a), and (4), a $T(z)$ is given by

(16) $$T(z) = F_A(z) F_B^{-1}(z).$$

In case (2b), a transformation is given by

(17) $$T(z) = F_A(z) \operatorname{diag}\{1, z^k\} \operatorname{diag}\{(\lambda_2 - \lambda_1)^{-k}, 1\} F_B^{-1}(z),$$

while in case (3b), a transformation is given by

(18) $$T(z) = F_A(z) \operatorname{diag}\{1, z^{-k}\} \operatorname{diag}\{(\lambda_1 - \lambda_2)^k, 1\} F_B^{-1}(z).$$

The transformation in (16) clearly has a formal power series expansion in z^{-1} with leading term I, while in (17) and (18), the series are formal Laurent series with not identically vanishing determinant and at most a finite number of terms with positive power of z. However, as a consequence of our theory, these particular quotients of (generally) divergent series do, in fact, converge. Since B(z) in (6) is given explicitly, it is easy to calculate as many terms in the above T(z) as we please, hence T(z) is effectively calculated.

We now obtain a fundamental solution for (3) as $\Phi(z) = T(z)\,\psi(z)$, where $\psi(z)$ is a fundamental solution for (6). Except in the trivial case (4), it is easy to transform a system of the form (6) into one which is equivalent to a confluent hypergeometric equation

$$w'' + (a\zeta^{-1} - 1)\,w' - b\zeta^{-1} w = 0.$$

For example, if $c \neq 0$ this is accomplished by means of the sequence of transformations

$$e^{\lambda_1 z}, \quad z^{\lambda'_1}, \quad \begin{bmatrix} 1 & 0 \\ 0 & zc^{-1} \end{bmatrix}, \quad \begin{bmatrix} 1 & 0 \\ dz^{-1} & 1 \end{bmatrix}, \quad z^d,$$

where d satisfies $d^2 - (\lambda'_2 - \lambda'_1)\, d - cc' = 0$, followed by the change of variable $z = (\lambda_2 - \lambda_1)^{-1}\, \zeta$ in the resulting second order (scalar) equation. Hence $\Psi(z)$ can be given explicitly in terms of elementary functions and confluent hypergeometric functions.

From the general theory, it is well-known that a fundamental matrix for (3) can be represented as

(;
$$\Phi(z) = S(z) z^M ,$$

where $S(z)$ is a 2×2 matrix of single-valued, analytic functions in $R < |z| < \infty$ with $\det S(z) \neq 0$, and M is a 2×2 constant matrix, called a <u>monodromy matrix</u> for the system. At an irregular singular point, there are no general methods which will produce M. Using the fact that systems (3) can be reduced to the form (6) by means of the above $T(z)$, and making use of μ introduced in (8), we explicitly give a Jordan canonical form for a monodromy matrix for (3) (and hence also for (1)) as follows :

If $\gamma = \gamma' = 0$, it is trivial from (15) that $M = \operatorname{diag}\{\lambda'_1, \lambda'_2\}$. Otherwise (i.e., at least one of γ, γ' is not zero), if 2μ is not an integer, then

$$M = \operatorname{diag}\left\{\mu + \tfrac{1}{2}(\lambda'_1 + \lambda'_2),\ -\mu + \tfrac{1}{2}(\lambda'_1 + \lambda'_2)\right\},$$

while if 2μ is an integer, then

$$M = \begin{bmatrix} \mu + \frac{1}{2}(\lambda'_1 + \lambda'_2) & 0 \\ 1 & \mu + \frac{1}{2}(\lambda'_1 + \lambda'_2) \end{bmatrix} .$$

The theory which underlies the above applications can be considered as an extension of the classical theory of canonical forms from matrices to linear systems of meromorphic differential equations. The transformations $T(z)$ which are considered are either analytic at ∞ with analytic inverse or are meromorphic at ∞ with not identically vanishing determinant. Such transformations are called, simply, analytic, resp., meromorphic, and the corresponding systems are called analytically, resp., meromorphically equivalent. Invariants are introduced, that is, quantities which remain unchanged with respect to (a type of) equivalence; for example, $\lambda_1, \lambda_2, \lambda'_1, \lambda'_2$, $\gamma\gamma'$ and μ are analytic invariants of (3). A collection of invariants is called complete if it characterizes the corresponding type of equivalence.

The main results in [5] concern two-dimensional systems $X' = A(z)\,X$, $A(z) = \sum_{i=0}^{\infty} A_i z^{-i}$, where A_0 has distinct eigenvalues and the power series converges for $|z|$ sufficiently large. We show how to compute a complete collection of invariants with respect to analytic and meromorphic equivalence by using a formal fundamental solution. In case of equivalence, we give an explicit (and constructive) description of all such transformations between two equivalent systems.

In the application above, the "matching" we carried out between the given system (3) and the special system (6) consisted of choosing the parameters in (6) so that its invariants coincided with those of (3). That this can be done is mainly a consequence of the fact that (6) can be solved (formally and actually) in terms of explicit integrals and infinite series, hence the invariants γ, γ' in (7) can be explicitly calculated in terms of the parameters in (6). Moreover, these expressions can be inverted [see (8)-(15)] to give the parameters c, c' in terms of γ, γ'. In all cases except (2b) and (3b), the matching was done using analytic invariants, however, in the cases (2b) and (3b), we used meromorphic invariants in the matching and consequently were led to the meromorphic transformations (17) and (18).

G. D. Birkhoff originated the idea of using transformations $T(z)$ to remove all unnecessary singularities from (3) and obtain a system of the form (6). He claimed in [1] that generally (i.e., for larger dimensional systems, higher order poles of $A(z)$ at ∞, and no assumption on the structure of A_0) by using analytic transformations a simplified system can be obtained whose coefficient has at most a simple pole at 0. Gantmacher ([4] ; 147), however, found a simple counter-example (two-dimensional system with a simple zero at ∞) to Birkhoff's statement. The cases (2b) and (3b) above are examples of systems which are not analytically equivalent to any system of the form (6), and therefore are also exceptions to Birkhoff's claim. In some of these cases, it can be shown directly (i.e., without using our theory) that all formal series beginning with a non-singular constant term and which transform an example of the form (2b) or (3b) into <u>any</u> system of the form (6) must diverge like $\sum n!\, z^{-n}$.

The main contribution from [1], which implies the existence of an analytic transformation that removes all finite, non-zero singularies from a simplified system and makes 0 at most a regular singular point, however, is valid and is used in the proof of our results, in particular, (7). Birkhoff[2] also discovered quantities related to the invariants and suggested some parts of the general program which we follow in order to obtain our results.

References

1. G. D. Birkhoff, "Equivalent singular points for ordinary linear differential equations," Math. Ann. 74 (1913), 134-139.

2. ______________, "On a simple type of irregular singular point," Trans. Amer. Math. Soc. 14(1913) 462-476.

3. E. Coddington and N. Levinson, Theory of Ordinary Differential Equations, McGraw-Hill, New York (1955).

4. F. R. Gantmacher, Theory of Matrices, vol. II, Chelsea, New York (1959).

5. W. Jurkat, D. Lutz, and A. Peyerimhoff, Birkhoff invariants and effective calculations for meromorphic linear differential equations, I (submitted).

Optimal Control of Limit Cycles or

What Control Theory can do to Cure a Heart Attack or to Cause One

Lawrence Markus

1. Control Theoretic Approach to Dynamical Systems

Control theory, as interpreted within the framework of dynamical systems or differential equations, leads to problems that are the inverses of the classical mathematical investigations. The classical theory of differential equations deals with analysis, whereas control theory deal with synthesis. In the classical approach to dynamical systems we are given the differential equations of motion, and then we try to analyse the behaviour of the resulting motions or solutions. In control theory we prescribe the desired behaviour of the solutions, and then we try to synthesize the differential equations to yield these motions. Of course, the procedure of synthesis means, in mathematical terms, that the basic form of the underlying differential equations can be modified by adjustment of certain control parameters or functional coefficients which are selected from certain admissible classes; whereas the synthesis means, in engineering terms, that the primary machine or plant can be modified by the adjustment of gains in feedback loops or the insertion of auxiliary devices of certain practical types.

Hence for each part of classical theory of differential equations, say stability or oscillation theory, there corresponds a field of control theory with inverse problems.

For instance, consider the classical stability analysis of the damped linear oscillator

$$\ddot{x} + 2b\dot{x} + k^2x = 0,$$

with constant coefficients. This oscillator is asymptotically stable (in the sense that all solutions approach $x = \dot{x} = 0$ as $t \to +\infty$) if and only if $b > 0$ and $k^2 > 0$. As an inverse problem assume $k > 0$ fixed and try to choose $b > 0$ so that the solutions are damped at the maximal rate. That is, define the cost or efficiency of the control parameter b to be

$$C(b) = \max \{ Re\lambda_1, Re\lambda_2 \} ,$$

where λ is any eigenvalue satisfying $\lambda^2 + 2b\lambda + k^2 = 0$.
Then we seek to select b to minimize $C(b)$.
An easy calculation shows that the optimal control b^* minimizing $C(b)$ is $b^* = k$. It is interesting to note that this is the standard value for critical damping, and hence we see that this familiar physical adjustment is explained as an elementary result in control theory.

As another illustration consider the forced oscillator

$$\ddot{x} + 2b\dot{x} + k^2 x = \sin \omega t ,$$

for constants $b > 0$, $k > 0$, $\omega > 0$. Classical analysis shows that there is a unique periodic solution

$$x = A \sin (\omega t + \phi)$$

with amplitude

$$A = [(k^2 - \omega^2)^2 + 4b^2\omega^2]^{-\frac{1}{2}} .$$

For the inverse problem fix $k > \sqrt{2}\, b > 0$ and try to choose the frequency $\omega > 0$ of the control input $\sin\omega t$ so as to maximize the amplitude $A(\omega)$ of the response output. An easy calculation shows that the optimal control ω^* maximizing $A(\omega) = [(k^2 - \omega^2)^2 + 4b^2 \omega^2]^{-\frac{1}{2}}$ is $\omega^* = (k^2 - 2b^2)^{\frac{1}{2}}$, which is assumed positive. Again we find this value familiar since ω^* is the resonating frequency, and hence this basic engineering tuning is explained as an elementary result in control theory.

These control theoretic results are interesting in that they illuminate well-known physical and engineering practice. Yet they are not typical of the modern theory of control. In the next section we comment on a standard formulation of modern control theory, and we develop an interesting generalization to a new class of problems concerning the control of a limit cycle.

2. Control of Limit Cycles and Applications to Cardiology

In control theory we consider a process or plant or dynamical system described by a differential system.

$$\dot{x} = f(x, u)$$

where x is the real state n-vector at time t, and the coefficient f is an n-vector function of the present state x and the control m-vector u. For simplicity we assume the process is autonomous (time-independent) and that f is continuous with continuous first derivatives for all $x \in R^n$ and $u \in R^m$, that is

$$f\colon\ R^n \times R^m \to R^n$$

is in class C^1.

We might seek to control $x(t)$ between given initial and final states in some fixed duration $0 \leq t \leq T$,

$$x(0) = x_0 \ , \quad x(T) = x_1 \ ,$$

by choosing a control function $u(t)$ from some admissible function class (say $u \in L_\infty[0, T]$, that is, $u(t)$ is a bounded measurable function on $0 \leq t \leq T$). Hence $x(t)$ is a solution of the two-point boundary value problem, with separated end conditions,

$$\dot{x} = f(x, u(t)) \ , \quad x(0) = x_0, \ x(T) = x_1 \ .$$

This constitutes the basic problem of controllability in control theory.

Among all solutions $x(t)$ to this boundary value problem, that is for all admissible control functions, we might try to select and describe the optimal solution $x^*(t)$ for the optimal controller $u^*(t)$ which minimizes some given cost or performance functional $C(u)$. This leads to the central problem of optimal control theory, for which there is a vast literature.

We now turn to a new type of control problem which corresponds to a boundary value problem with periodic conditions rather than separated conditions. That is, we again consider the plant or control dynamics

$$\dot{x} = f(x, u) ,$$

but the admissible controllers $u(t)$ are now required to be periodic functions, with some period T. Under certain assumptions, each periodic controller $u(t)$ leads to a periodic response $x(t)$ of

$$\dot{x} = f(x, u(t)) ,$$

where $x(t) \equiv x(t + T)$ for all real t. That is, $x(t)$ is a periodic solution or, in the terminology of oscillation theory, $x(t)$ is a limit cycle. We do not prescribe any initial or final states except for the periodicity boundary condition

$$x(0) = x(T)$$

We might proceed further to select and describe that limit cycle $x^*(t)$, which minimizes or maximizes some performance functional $C(u)$. In particular, we shall let $C(u)$ be the amplitude of the limit cycle $x(t)$, and find the optional controller $u^*(t)$ to maximize this amplitude.

From the purely mathematical point of view more general types of boundary conditions could be utilized, or the periodic functions could be generalized to almost periodic functions.

While various extensions of our theory could be pursued, we shall concentrate on the control of a limit cycle. This mathematical problem was motivated by some engineering instrumentation of cardiac assist devices related to heart surgery. For instance, a heart pump must be designed to assist the heartmaintain its natural amplitude of systolic and diastolic pressures. Further design improvements would force the controlled heart to maintain a circulatory regime very near to the natural healthy action. Unfortunately the dynamics of the human circulatory system are known too poorly for a useful application of any very sophisticated mathematical or engineering theory. Thus this study can be considered as an introduction to a developing engineering-medical field that could become of great practical significance.

3. Linear Dynamics: Geometry of Limit Cycle Control

We consider a linear control system

$$\dot{x} = Ax + Bu$$

where the state x is a real (column) n-vector, the control u is a real m-vector, and A and B are real constant matrices. For each bounded measurable control u(t), we have the solution x(t) initiating at the state x_0 as prescribed by the Lagrange formula of variations of parameters,

$$x(t) = e^{At}x_0 + e^{At}\int_0^t e^{-As}\,Bu(s)\,ds\ .$$

If u(t) is periodic, say $u(t) \equiv u(t+T)$ for almost all real times t, then x(t) has the same period T just in case x(T) = x(0), that is,

$$x_0 = e^{AT}x_0 + e^{AT}\int_0^T d^{-As}\,Bu(s)\,ds\ .$$

Hence there exists a unique response x(t) with period T, and this initiates at

$$x_0 = (I-e^{AT})^{-1}\,e^{AT}\int_0^T e^{-As}\,Bu(s)\,ds\ ,$$

provided the matrix $(I-e^{AT})$ is invertible. We shall assume, for simplicity, that T=1 and the matrix A has no pure imaginary eigenvalues, $\operatorname{Re}\lambda(A) \neq 0$.

Then $I-e^A$ is invertible, and the initial point $x_0[u]$ for the unique periodic response $x(t)$ to the periodic control $u(t)$ is given by

$$x_0[u] = (I-e^A)^{-1} e^A \int_0^1 e^{-As} Bu(s)\, ds .$$

The probelm of optimal control of the amplitude of the limit cycle now depends on finding the admissible controller $u^*(t)$ for which $x_0[u^*]$ leads to the periodic response $x^*(t)$ having the maximal amplitude. We shall restrain the periodic control inputs $u(t)$ by the condition

$$u(t) \in \Omega \quad \text{for all } t,$$

where Ω is a given compact convex subset of the real m-vector space R^m. For instance, Ω could be the cube of unit radius centred at the origin, say $|u^i| \le 1$ for $i = 1, \ldots, m$. Hence an admissible controller $u(t)$ is a measurable vector in Ω having period of one.

The amplitude of the response $x(t)$ will be taken to be the maximum value of the first component $x^1(t)$ of the vector $x(t)$. That is

$$C(u) = -\max_{0 \le t \le T} x^1(t)$$

and we seek to minimize the cost $C(u)$. Of course, other costs could be studied, for instance the norm

$$\|x(t)\| = \max_{0 \le t \le T} [\,|x^1(t)| + \cdots + |x^n(t)|\,] .$$

<u>Definition.</u> Consider the linear control dynamics in R^n

$$\dot{x} = Ax + Bu$$

with compact convex restraint set $\Omega \subset R^m$. For each measurable controller $u(t) \in \Omega$ with period 1, there is a unique initial point $x_0[u]$ for the response of period 1 (assuming A has no pure imaginary eigenvalues). Define the subset of R^n

$K = \{ x_0[u] \mid \text{for all admissible controllers } u(t) \}$.

The set K corresponds to the attainable set in the usual controllability problem (separated boundary conditions), and the letter K korresponds to the konditions on kompactness and konvexity

Theorem 1. Consider the linear control dynamics in R^n

$$\dot{x} = Ax + Bu \qquad (\mathrm{Re}\lambda(A) \neq 0)$$

with compact convex restraint set $\Omega \subset R^m$.

Let $K = \{ x_0[u] \mid u(t) \equiv u(t+1) \in \Omega \text{ almost all } t \in R^1 \}$.

Then

i) K is a compact convex subset of R^n, and

ii) K is the union of all periodic responses $x(t)$.

Proof.

Consider the map

$$u \to x_0[u] = (I-e^A)^{-1} e^A \int_0^1 e^{-As} Bu(s)\, ds.$$

The set of admissible control functions $\mathcal{J}$ is convex, since Ω is convex. Also the map $u \to x_0[u]$ is linear, and so K is convex.

The function set $\mathcal{J}$ can be embedded in some closed ball B_2 of the Hilbert space $L_2[0,1]$. We recall that B_2 is weakly compact and any sequence $u_k \in \mathcal{J}$ has a weakly convergent subsequence

$$u_{k_i} \longrightarrow u^* .$$

Consider points $x_0[u_k] \in K$, and select a subsequence corresponding to controllers u_{k_i} converging weakly to u^*. If $u^* \in \mathcal{J}$, then it follows that

$x_0[u_{k_i}] \to x_0[u^*] \in K$.

Of course, we can define $u^*(t) \in L_2[0,1]$ to have period one on R^1. But we must still verify that the values of $u^*(t)$ lie in Ω, for almost all time t. Recall that the weak limit of positive functions is positive (almost everywhere), and use this result to conclude that $u^*(t)$ lies a.e. in any half-space of R^n that contains Ω. But Ω is the intersection of a countable number of closed half-spaces, and hence $u^*(t)$ lies in Ω almost everywhere for $t \in R^1$.

Thus $u^* \in \mathcal{J}$ is an admissible controller, and $x_0[u^*] \in K$. Hence K is compact, and the first conclusion of the theorem has been demonstrated.

The second conclusion follows from the autonomous nature of the control dynamics. For let $u(t) \in \mathcal{J}$ yield the initial state $x_0[u]$ on the periodic response $x(t)$. Then, for each positive number τ, the controller $u(t+\tau)$ yields the periodic response $x(t+\tau)$ with initial state $x(0+\tau)$. Thus the entire set $\{x(t) \mid 0 \leq t \leq 1\}$ lies within K, as required. Q.E.D.

Remark 1. The crux of the above proof concerns the linear map $u \to x_0[u]$ of $\mathcal{J} \to R^n$. If $\mathcal{J}$ is a convex weakly compact (more accurately, weak star sequentially compact) set, then the image K is convex and compact. Thus if $\mathcal{J} = B_p$, a closed ball in the Banach space $L_p[0,1]$ for some $1 < p \leq \infty$, then K is convex and compact.

Remark 2. It is also possible to draw the conclusions of theorem 1 with the assumption that Ω is compact, but not necessarily convex in R^m. Recall that the set $\{ e^A \int_0^1 e^{-As} Bu(s)ds \mid u(s) \in \Omega \}$ is the set of attainability from the origin in unit time, and is consequently compact and convex [1].
But this set, when multiplied by the nonsingular matrix $(I-e^A)^{-1}$, is precisely K.

In the problem of controllability from the origin, we recall that set of attainability is a convex body (that is, has nonempty interior) in case Ω contains a neighorhood of $u=0$ in R^m, and the controllability condition holds

$$\text{rank } [B, AB, A^2B, \ldots, A^{n-1}B] = n.$$

This yields the following result, see [1 p. 84].

Corollary. Let Ω be a compact convex neighborhood of the origin in R^m. Then K is a convex body containing the origin in its interior in R^n just in case

$$\text{rank } [B, AB, A^2B, \ldots, A^{n-1}B] = n.$$

Remark 3. It is sometimes of interest to remove the restraint Ω on controllers. In this case the linearity of the dynamics and the above corollary assert that the set

$$K_\infty = \{x_0[u] \mid u(t) \equiv u(t+1) \text{ in } L_\infty(-\infty, \infty)\}$$

is a linear subspace of R^n. Moreover $K_\infty = R^n$ just in case the controllability condition holds

$$\text{rank } [B, AB, A^2B, \ldots, A^{n-1}B] = n.$$

We next turn to an examination of the boundary ∂K of the set $K \subset R^n$, and we shall prove the appropriate form of the maximal principle for our problem.

Theorem 2. Consider the linear control dynamics in R^n

$$\dot{x} = Ax + Bu \qquad (\text{Re}\lambda(A) \neq 0)$$

with compact convex restraint set $\Omega \subset R^m$. Then an admissible controller $u^*(t+1) \equiv u^*(t)$ (a. e.) in Ω yields a point $x_0[u^*] \in \partial K$ if and only if:

i) there exists a nontrivial solution $\eta^*(t)$ of $\dot{\eta} = -\eta A$ which satisfies the maximal principle

ii) $\eta^*(t)Bu^*(t) = \max_{u \in \Omega} \eta^*(t)Bu$, almost always.

Proof. Let $u^*(t)$ be an admissible controller such that

$$x_0[u^*] = (I-e^A)^{-1} e^A \int_0^1 e^{-As} Bu^*(s)ds \in \partial K.$$

Since K is a compact convex set in R^n, there exists a unit exterior vector η^* such that

$\eta^* x_0[u^*] \geq \eta^* x_0$, for all points $x_0 \in K$.

We now define the nontrivial row vector

$$\eta^*(t) = \eta^* (I-e^A)^{-1} e^A e^{-At}$$

as a solution of the linear system $\dot{\eta} = - \eta A$.

The inequality or maximal principle satisfied by $\eta^* x_0[u^*]$ yields the results,

$$\int_0^1 \eta^*(s) \; Bu^*(s) \, ds \geq \int_0^1 \eta^*(s) \; Bu(s) \, ds ,$$

for all admissible controllers $u(s) \subset \Omega$. But this implies that

$$\eta^*(s) \; Bu^*(s) = \max_{u \in \Omega} \eta^*(s) \; Bu, \quad \text{almost always on} \quad 0 \leq s \leq 1.$$

For otherwise,

$$\eta^*(s) \; Bu^*(s) < \max_{u \in \Omega} \eta^*(s) \; Bu$$

on some positive duration Σ. In this case we could define the admissible controller $\hat{u}(s)$ such that

$$\eta^*(s) \; B\hat{u}(s) = \max_{u \in \Omega} \eta^*(s) \; Bu \quad \text{for} \quad s \in \Sigma,$$

and

$$\hat{u}(s) = u^*(s) \qquad \text{for} \quad s \in [0,1] - \Sigma .$$

With this controller $\hat{u}(s)$ we compute

$$\int_0^1 \eta^*(s) \; Bu^*(s) \, ds < \int_0^1 \eta^*(s) \; B\hat{u}(s) \, ds ,$$

which contradicts the earlier inequality. Therefore the maximal principle holds

$$\eta^*(s) \; Bu^*(s) = \max_{u \in \Omega} \eta^*(s) \; Bu , \quad \text{almost always.}$$

Conversely, let the controller $u^*(t)$ with response $x^*(t)$, and a nontrivial solution $\eta^*(t)$ of the adjoint differential system $\dot{\eta} = - \eta A$ all satisfy the maximal principle

$$\eta^*(t)\, Bu^*(t) = \max_{u \,\varepsilon\, \Omega} \eta^*(t)\, Bu, \quad \text{almost always.}$$

We must show that $x^*(0) = x_0[u^*] \;\varepsilon\; \partial K$.

Suppose the contrary, that $x_0[u^*]$ were an interior point of K in R^n. Then for every nonzero vector η_0, there would exist an interior point $x_0 \;\varepsilon\; K$ for which

$$\eta_0\, x^*(0) < \eta_0\, x_0 .$$

Now let us define η_0 by $\eta_0 = \eta^*(1)\,(I-e^A)$ so

$$\eta^*(t) = \eta_0 (I-e^A)^{-1}\, e^A\, e^{-At} .$$

Then the maximal principle asserts that

$$\int_0^1 \eta_0 (I-e^A)^{-1}\, e^A\, e^{-As}\, Bu^*(s)\, ds \geq \int_0^1 \eta_0 (I-e^A)^{-1}\, e^A e^{-As} Bu(s)\, ds,$$

with strict inequality unless $u(s)$ satisfies the maximal principle.
This means that

$$\eta_0\, x_0\,[u^*] \geq \eta_0\, x_0\,[u] ,$$

for all points $x_0[u] \;\varepsilon\; K$. But this contradicts the supposition that $x_0[u^*]$ lies interior to K. Hence $x_0[u^*] \;\varepsilon\; \partial K$, as required. Q.E.D.

From the boundary position of $x_0[u^*]$, we can often deduce that $u^*(t)$ lies on the boundary of Ω. The resulting behaviour of $u^*(t)$ is known as bang-bang control, as discussed in the next theorem. We shall require an algebraic hypothesis, the normality condition, which is stronger than the controllability condition. Also we shall demand that the restraint set Ω is a convex polytope, that is, the intersection of a finite number of closed half-spaces of R^m (and Ω is not just a single point).

<u>Theorem 3</u>. <u>Consider the linear control dynamics in</u> R^n

$$\dot{x} = Ax + Bu \qquad (\mathrm{Re}\lambda(A) \neq 0)$$

with restraint set Ω that is a convex polytope in R^m.

Assume the normality condition:

$$Bw,\ ABw,\ A^2Bw,\ \ldots,\ A^{n-1}\ Bw$$

are linearly independent for every vector w which is an edge of Ω (or along the segment Ω if $m = 1$). Then each controller $u^*(t)$ with $x^*(0) \varepsilon \partial K$ is a bang-bang controller, that is $u^*(t)$ is piecewise constant with values only at the vertices of Ω, and with only a finite number of switches.

Proof. Let $u^*(t)$ yield the point $x_0[u^*] = x^*(0) \ \varepsilon \ \partial K$. Then by the maximal principle there is a nontrivial solution

$$\eta^*(t) = \eta^* (I - e^A)^{-1} e^A e^{-At} \quad \text{of } \dot{\eta} = - \eta A \text{ such that}$$

$$\eta^*(t) = Bu^*(t) = \max_{u \varepsilon \Omega} \eta^*(t) Bu \text{ , almost everywhere.}$$

Here η^* is a unit exterior normal to a supporting hyperplane to K at $x^*(0) \varepsilon \partial K$.

Consider, at each instant t, the real linear function of u,

$$F_t(u) = \eta^*(t) Bu \qquad \text{for } u \varepsilon \Omega .$$

Since Ω is a convex polytope in R^m, the linear function $F_t(u)$ assumes its maximum where u lies on a face of Ω (or edge, or vertex of Ω) and on this face it is constant.

Let S be the set of all times when $F_t(u)$ has its maximum on edges of Ω. Suppose S has infinitely many points. Since Ω has only finitely many edges, there would then be an infinite set S_1 during which $F_t(u)$ takes its maximum on a fixed edge e_1 of Ω. Let w be a unit vector along the edge e_1, and then

$$\eta^*(t) Bw = 0 \qquad \text{for all } t \ \varepsilon \ S_1.$$

But a real analytic function on $0 \leq t \leq 1$, which vanishes on an infinite set S_1, must be identically zero. Thus

$$\eta^* (t) Bw \equiv 0 \qquad \text{for } 0 \leq t \leq 1.$$

We write $\eta^*(t) = \eta_0 e^{-At}$ where $\eta_0 = \eta^*(I-e^A)^{-1} e^A$ is a nonsingular constant initial vector, and differentiate

$\eta_0 e^{-At} Bw \equiv 0$ to get $\eta_0 Ae^{-At} Bw \equiv 0$ so $\eta_0 ABw = 0$.

Continue to differentiate and evaluate at $t = 0$ to get

$\eta_0 Bw = 0, \eta_0 ABw = 0, \eta_0 A^2 Bw = 0, \ldots, \eta_0 A^{n-1} Bw = 0.$

But this implies that the n different vectors $Bw, ABw, \ldots, A^{n-1} Bw$ are all orthogonal to η_0, and hence they must be linearly dependent. This contradicts the normality hypothesis, and so we see that $F_t(u)$ assumes its maximum only at vertices of Ω (excepting a finite set of times). Also, $F_t(u)$ switches between vertices only when $t \in S$.

Now $u^*(t)$ satisfies the maximal principle

$$\eta^*(t) Bu^*(t) = \max_{u \in \Omega} F_t(u), \quad \text{almost always,}$$

Thus $u^*(t)$ must lie at the vertex of Ω which is the single point of Ω where $F_t(u)$ takes its maximum value. That is, after possible re-definition on a null set of time, $u^*(t)$ is a bang-bang controller taking values only at the vertices of Ω, and switching between these vertices only finitely many times. Q.E.D.

Corollary. Let Ω be a convex polytope and assume the normality condition holds. Then K is a strictly convex body, and each point $x_0 \in \partial K$ is achieved by a unique controller.

Proof. Suppose K were not strictly convex in R^n. Then points $x_0[u_1] \neq x_0[u_2]$ would lie on one supporting hyperplane to K. Let η^* be a corresponding exterior normal vector to K, and consider

$\eta^*(t) = \eta^*(I-e^A)^{-1} e^A e^{-At}$ as a solution of $\dot{\eta} = -\eta A$.

But $u_1(t)$ and $u_2(t)$ each satisfy the maximum principle with $\eta^*(t)$,

$$\eta^*(t)\, Bu_1(t) = \eta^*(t)\, Bu_2(t) = \max_{u \in \Omega} \eta^*(t)\, Bu, \quad \text{a.e.}$$

From the proof of theorem 3 we find that

$$u_1(t) = u_2(t) \quad \text{almost always on } 0 \leq t \leq 1,$$

and this implies that $x_0[u_1] = x_0[u_2]$. Thus K is strictly convex, and hence is a convex body in R^n.

The above argument also shows that if controllers $u_1(t)$ and $u_2(t)$ yield the same point $x_0 \in \partial K$, then $u_1(t) = u_2(t)$, almost everywhere. Q.E.D.

4. Optimal Control of Amplitudes of Limit Cycles for Linear Dynamics

We shall consider a linear dynamical system

$$\dot{x} = Ax + Bu \qquad (\mathrm{Re}\,\lambda(A) \neq 0)$$

with the state $x(t)$ in R^n, and the periodic control $u(t) \equiv u(t+1)$ in a compact convex restraint set Ω in R^m. For each admissible controller (bounded measurable m-vector function $u(t) \in \Omega$ and with period 1), there exists a unique periodic response $x(t)$ initiating from $x(0) = x_0[u]$ in the set K. The cost of the control $u(t)$ is given by the functional

$$C(u) = - \max_{0 \leq t \leq 1} x^1(t),$$

where x^1 is the first component of the state x.
We seek to determine an optimal controller $u^*(t)$ which minimizes $C(u)$, so

$$C(u^*) \leq C(u) \qquad \text{for all admissible } u(t).$$

If x^1 measures some physical displacement, then $\max_{0 \leq t \leq 1} x^1(t)$ is a measure of the amplitude of the state oscillation, and $u^*(t)$ maximizes the amplitude. Other definitions of the state amplitude are possible, but the above choice is particularly easy to study.

Theorem 4. Consider the linear control dynamics in R^n

$$\dot{x} = Ax + Bu \qquad (\mathrm{Re}\lambda(A) \neq 0)$$

with compact convex restraint set $\Omega \subset R^m$.

For each admissible controller $u(t) \equiv u(t+1) \varepsilon \Omega$ a. e, let the cost be

$$C(u) = - \max_{0 \leq t \leq 1} x^1(t),$$

where $x^1(t)$ is the first component of the periodic response $x(t)$.

Then i) there exists an optimal controller $u^*(t)$ with $u^*(t)$ satisfying the the maximal principle

ii) $\eta^*(t) Bu^*(t) = \max_{u \varepsilon \Omega} \eta^*(t) Bu$, a. e., where

$$\eta^*(t) = (1, 0, 0, \ldots, 0) (I-e^A)^{-1} e^A e^{-At} .$$

Furthermore assume that Ω is a convex polytope and the normality condition holds :

$$Bw, ABw, A^2Bw, \ldots, A^{n-1} Bw$$

are linearly independent for each edge w of Ω.

Then, up to time translations and discarding null time-durations,

iii) $u^*(t)$ is the unique optimal controller, and

iv) $u^*(t)$ is a bang-bang controller over the vertices of Ω having a finite number of switches.

Proof. The set K of all initial points $x_0[u]$ of admissible controllers is a compact convex set, and thereon the real function x^1 assumes its maximum at some point $x_0^* \varepsilon \partial K$. Let $\pi : x^1 = x_0^{*1}$ be the supporting hyperplane to K at x_0^* and let $\eta^* = (1, 0, 0, \ldots, 0)$ be the exterior normal to π. Since K is the union of all periodic responses, no periodic response $x(t)$ has a first component that ever exceeds x_0^{*1}. Thus the periodic response $x^*(t)$

initiating at x_0^* is an optional response and this corresponds to some optimal controller $u^*(t)$ for which $x_0[u^*] = x_0^*$

By theorem 2 we note that $u^*(t)$ satisfies the maximal principle

$$\eta^*(t)\, Bu^*(t) = \max_{u \, \varepsilon \, \Omega} \eta^*(t)\, Bu \qquad \text{a. e.,}$$

where

$$\eta^*(t) = \eta^*(I-e^A)^{-1}\, e^A\, e^{-At},$$

as required above.

Now assume that Ω is a convex polytope and that the normality condition holds. Then K is a strictly covex body, so the supporting hyperplane π meets in the single point x_0^*. Let $u_1(t)$ be an optimal control with periodic response $x_1(t)$. Then, at some time τ, $x_1(\tau) = x_0^*$. But this means that $u_1(t+\tau)$ yields the response $x_1(t+\tau)$ initiating at x_0^*. Since there is a unique controller $u^*(t)$ for which the periodic response initiates at x_0^*, we conclude that $u_1(t+\tau) = u^*(t)$.

Hence $u^*(t)$ is the unique optimal controller, up to time-translations and time null sets. By theorem 2 $u^*(t)$ is a bang-bang controller, as required.

Q. E. D.

We next present an example of optimal control of limit cycle. It is interesting to compare this example with the classical calculation of the resonance frequency of a linear oscillator, as discussed in the first section of this paper above.

Example. Consider the linear oscillator

$$\ddot{x} + 2b\,\dot{x} + k^2\,x = u(t)$$

where the controller has period 1 and satisfies the restraint $|u(t)| \leq 1$. The constant coefficients satisfy $k > b > 0$ so $\zeta = (k^2 - b^2)^{\frac{1}{2}} > 0$. We seek to find the optimal controller $u^*(t)$ whose periodic response $\begin{pmatrix} x^*(t) \\ \dot{x}^*(t) \end{pmatrix}$ has the maximum amplitude $\max_{0 \leq t \leq 1} x^*(t)$.

In the phase plane R^2 the dynamical system is

$$\begin{pmatrix} \dot{x} \\ \dot{y} \end{pmatrix} = A \begin{pmatrix} x \\ y \end{pmatrix} + Bu,$$

where

$$A = \begin{pmatrix} 0 & 1 \\ -k^2 & -2b \end{pmatrix}, \qquad B = \begin{pmatrix} 0 \\ 1 \end{pmatrix}$$

and Ω is the segment $-1 \leq u \leq 1$ in R^1. Clearly the normality condition is satisfied, since

$$(Bw, \ ABw) = \begin{pmatrix} 0 & 1 \\ 1 & -2b \end{pmatrix}, \quad \text{where } w = (1),$$

has rank 2.

Then the unique optimal controller $u^*(t)$ is determined by the maximal condition

$$\eta^*(t) \begin{bmatrix} 0 \\ 1 \end{bmatrix} u^*(t) = \max_{|u| \leq 1} \eta^*(t) \begin{bmatrix} 0 \\ 1 \end{bmatrix} u .$$

Thus

$$u^*(t) = \operatorname{sgn} \eta_2(t) \qquad \text{a. e.},$$

where

$$(\eta_1(t), \eta_2(t)) = \eta^*(t) = (1, 0)\,(I - e^A)^{-1} e^A \, e^{-At} .$$

Use the exponential matrix

$$e^{At} = \frac{k}{\zeta} e^{-bt} \begin{pmatrix} \sin(\zeta t + \alpha) & , \ \frac{1}{k} \sin \zeta t \\ -b \sin(\zeta t + \alpha) + \zeta \cos(\zeta t + \alpha), & \frac{-b}{k} \sin \zeta t + \frac{\zeta}{k} \cos \zeta t \end{pmatrix}$$

where $\zeta = (k^2 - b^2)^{\frac{1}{2}}$, $\sin \alpha = \frac{\zeta}{k}$, $\cos \alpha = \frac{b}{k}$,

and compute the optimal controller

$$u^*(t) = \operatorname{sgn} \frac{e^{-b+2bt}}{\Delta} \left[-\frac{1}{k} \sin \zeta(1-t) + \frac{\zeta}{k^2} \sin \zeta t \right],$$

where $\Delta = 1 - \frac{2\zeta}{k} \cos \zeta + \frac{\zeta^2}{k^2}$.

Assuming $b << k$ so $\Delta > 0$, we can write

$$u^*(t) = \operatorname{sgn}\left[\sin(\zeta t - \zeta) + \frac{\zeta}{k} \sin \zeta t\right].$$

This formula appears strange since various dimensionality factors (period, control restraint) have been normalized at unity. Also this optimal controller yields a periodic response $x^*(t)$ which realizes its maximal amplitude $x^*(0)$ at the initial instant $t = 0$. We note that there is no relation between the chosen control period of 1, and the motion of the free oscillator (which, in fact, has no nontrivial periodic solutions).

We do not evaluate the maximal amplitude $x^*(0)$, which can best be done numerically,

5. Nonlinear Dynamics: Control of Limit Cycles

Consider a nonlinear control system in R^n

$$\dot{x} = f(x, u)$$

where $u(t) = u(t+1)$ is a periodic control vector lying in a compact set $\Omega \subset R^m$. Suppose there exists a unique periodic response $x(t) = x(t+1)$ lying in some (often) compact constraint set $\Lambda \subset R^n$. We shall seek an optimal controller $u^*(t)$ for which the response $x^*(t)$ assumes the maximal amplitude, in the sense that the first component $x^{*1}(t)$ achieves the maximum possible value at $t = 1$.

Examples from the literature on nonlinear vibrations are illustrative. For instance consider the scalar oscillator

$$\ddot{x} + \dot{x} + x + x^3 = u(t) \quad .$$

In the (x, y) phase plane R^2 this system becomes

$$\dot{x} = y$$
$$\dot{y} = -x - x^3 - y + u(t) \quad ,$$

and we seek periodic solutions $(x(t), y(t))$ in a small disc $\Lambda: x^2 + y^2 \leq \rho^2$ when the periodic control $u(t)$ is restricted to a small interval $\Omega: |u| \leq \alpha$. The general theory of perturbations assures us that for a fixed small $\rho > 0$ there exists suitably small $\alpha > 0$ such that each 1-periodic controller in Ω produces a unique 1-periodic response in Λ.

A more general type of controlled oscillator is described by the vector system in R^n

$$\dot{x} = f(x) + Bu(t) \quad ,$$

for a constant nxm matrix B. Assume that some basic equation (with $u=u_0(t)$) has a periodic solution $x = \varphi(t)$ with $\varphi(t) = \varphi(t+1)$ giving the shortest period. Also assume that this periodic solution of the free equation is stable, or at least that the Poincaré map

$$x_0 \quad \rightarrow x(1, x_0)$$

(where $x(t, x_0)$ is the solution of $x = f(x) + Bu_0(t)$ initiating at x_0) does not have 1 as a characteristic multiplier, that is the matrix $P = \frac{\partial x}{\partial x_0} (1, \varphi(0))$ has no eigenvalue 1. Then we fix a compact tubular neighborhood Λ of $\{ x = \varphi(t) \}$ and thereafter a small closed neighborhood Ω of $\{ u = u_0(t) \}$. The general theory of perturbations assures us that each 1-periodic controller in Ω produces a unique 1-periodic response in Λ.

Theorem 5. *Consider the control system in* R^n

$$\dot{x} = f(x,u) = f(x) + B(x)\,u$$

where $f(x,u) \in C^1$ *for* (x,u) *in a compact set* $\Lambda \times \Omega \subset R^n \times R^m$ *and* Ω *is convex.*

Assume that each measurable controller

$$u(t) = u(t+1) \;\in\; \Omega \;\; (a.e)$$

yields some 1-periodic response $x_u(t) \in \Lambda$. *Let the cost of* $u(t)$ *be given by*

$$C(u) = g[x_u] + \int_0^1 (f^0(x_u) + G(x_u)u)\,dt,$$

where g *is a continuous functional on the function space* $C[0,1]$ *and* $f^0(x,u) = f^0(x) + G(x)\,u$ *is continuous in* R^{n+m}.

Then there exists an optimal controller $u^*(t)$ *minimizing (or maximizing) the cost at* $C(u^*)$.

Proof. Since Ω and Λ are compact, and $f^0(x,u)$ and g are continuous, we see that inf $C(u) = m$ is finite. Take a sequence $u^{(k)}(t)$ of controllers, with corresponding responses $x^{(k)}(t)$, so $C(u^{(k)}) \searrow m$. Using subsequences we can assume (with usual arguments such as those in [1] p.260)

$$u^{(k)}(t) \rightharpoonup u^*(t) \qquad \text{weakly in } L_2[0,1].$$

$$x^{(k)}(t) \rightrightarrows x^*(t) \qquad \text{uniformly in } C[0,1]$$

and also

$$f(x^{(k)}(t), u^{(k)}(t)) \rightharpoonup f(x^*(t), u^*(t))$$

$$f^0(x^{(k)}(t), u^{(k)}(t)) \rightharpoonup f^0(x^*(t), u^*(t)) \qquad .$$

Now $x^{(k)}(t)$ is the response to $u^{(k)}(t)$ so

$$x^{(k)}(t) = x^{(k)}(0) + \int_0^t [f(x^{(k)}(s)) + B(x^{(k)}(s))\, u^{(k)}(s)]\, ds$$

and

$$x^*(t) = x^*(0) + \int_0^t [f(x^*(s)) + B(x^*(s))\, u^*(s)]\, ds.$$

Thus $x^*(t) = x^*(t+1) \in \Lambda$ is a periodic response to the controller $u^*(t)$. Using the convexity of Ω we can show easily that $u^*(t) \in \Omega$ is an admissible controller.

Further calculations with the cost functional yield

$$C(u^{(k)}) = g\,[x^{(k)}] + \int_0^1 [f^0(x^{(k)}) + G(x^{(k)})\, u^{(k)}(t)]\, dt.$$

As $k \to \infty$ we compute $C(u^{(k)}) \searrow m$ and

$$\lim_{k \to \infty} C(u^{(k)}) = g[x^*] + \int_0^1 [f^0(x^*) + G(x^*)\, u^*(t)]\, dt .$$

Thus

$$C(u^*) = m ,$$

and $u^*(t)$ is an optimal controller, as required. Q.E.D.

Remark. If we define the cost functional

$$C(u) = g[x] = - \max_{0 \le t \le 1} x^1(t) ,$$

then an optimal controller $u^*(t)$ that minimizes $C(u)$ will maximize the amplitude, in the above sense.

We next turn to the maximal principle as a necessary condition for an optimal control of the amplitude of a limit cycle for a nonlinear dynamical system. The concepts and methods are similar to those used in the standard formulation of the maximal principle for the control of a trajectory between given endpoints, see [1] pp. 246-256. We use the notations and calculations of this text with no further explanation.

Theorem 6. Consider the control system in R^n

$$\dot{x} = f(x, u) \text{ in } C^1 \text{ in } R^{n+m}.$$

Use all measurable controllers $u(t) = u(t+1)$ lying in the compact restraint set $\Omega \subset R^m$ (a.e.), some of which have 1-periodic responses $x(t)$ in R^n. Let $u^*(t)$ be an optimal controller with response $x^*(t)$ maximizing the amplitude $x^{*1}(t)$ at $x^{*1}(1)$.

Assume that $x^*(t)$ has a Poincaré map such that

$$P = \frac{\partial x}{\partial x_0}(t, x_0) \Bigg|_{\substack{t=1 \\ x_0 = x^*(e)}} \text{ has no eigenvalue of } 1$$

(where $x(t, x_0)$ is general solution of $\dot{x} = f(x, u^*(t))$ initiating at x_0).

Then there exists a nontrivial row n-vector $\eta^*(t)$ satisfying the adjoint variational equation

i) $$\dot{\eta} = -\eta \frac{\partial f}{\partial x}(x^*(t), u^*(t)),$$

and the maximal principle

ii) $$\eta^*(t) f(x^*(t), u^*(t)) = \max_{u \in \Omega} \eta^*(t) f(x^*(t), u) \quad \text{(a.e.)}$$

with the terminal condition

iii) $$\eta^*(1) = (1, 0, 0, \dots, 0)(I - P)^{-1}.$$

Remarks. Before discussing the proof of the maximal principle we consider the special case of an autonomous linear system in R^n

$$\dot{x} = Ax + Bu$$

Here $\eta^*(t)$ satisfies

$$\dot{\eta} = -\eta A, \quad \text{since} \quad A = \frac{\partial f}{\partial x} ,$$

and the maximal principle becomes

$$\eta^*(t)\,[Ax^*(t) + Bu^*(t)] = \max_{u\,\varepsilon\,\Omega} \eta^*(t)\,[Ax^*(t) + Bu] ,$$

or more simply

$$\eta^*(t)\,Bu^*(t) = \max_{u\,\varepsilon\,\Omega} \eta^*(t)\,Bu .$$

The terminal condition on $\eta^*(t)$ is found by computing $P = e^A$ from the formula

$$x(t, x_0) = e^{At} x_0 + \int_0^t e^{A(t-s)} Bu^*(s)ds .$$

Clearly P has no eigenvalue 1 just in case A has no eigenvalue that is an integral multiple of the pure imaginary $2\pi i$. In this situation the above theorem on the maximal principle reduces to earlier results for linear systems.

Just as for the earlier linear analysis, the general form of the maximal principle for nonlinear dynamics can often be used to display the bang-bang character of the optimal controller $u^*(t)$ and to guide in the computation of $u^*(t)$.

6. Proof of Maximal Principle

We shall only sketch the main ideas in the proof of the theorem, the details following the pattern offered in [1].

If $x^*(t)$ is an optimal response, with $x^{*1}(1)$ achieving the maximal amplitude, then

$$(1, 0, 0, \ldots, 0)\; x^*(1) = \max\, (1, 0, 0, \ldots, 0)\; x\,(1),$$

where $x(t)$ is any 1-periodic response to any admissible controller $u(t)$. In other words, the hyperplane $x^1 = x^{*1}(1)$ contains all the possible points $x(1)$

on one side. This is the basic geometric fact which yields the maximal principle as an analytical interpretation.

We shall perturb $u^*(t)$ to a new admissible controller $u_\pi(t, \varepsilon)$ with periodic response $u^*_\pi(t, \varepsilon)$ (as defined below) and for $\varepsilon = 0$ the perturbed controller and response reduce to $u^*(t)$ and $x^*(t)$. Then we can assert that

$$(1, 0, 0, \dots, 0)\, x^*(1) \geq (1, 0, 0, \dots, 0)\, x^*_\pi(1, \varepsilon)$$

for every perturbation π.

We approximate $x^*_\pi(1, \varepsilon) - x^*(1)$ by a vector $\frac{\partial x^*_\pi}{\partial \varepsilon}(1, 0)$ based at $x^*(1)$, and write the maximal principle as

$$-(1, 0, 0, \dots, 0)\, \frac{\partial x^*_\pi}{\partial \varepsilon}(1, 0) \geq 0 .$$

Our approximation and the resulting conclusions will be valid for suitably small $\varepsilon > 0$.

We shall define a perturbation by data $\pi = \{t_1, \ell_1, u_1\}$ where t_1 is an instant on $0 \leq t \leq 1$, $\ell_1 \geq 0$, and u_1 is an arbitrary point in the set Ω. Essentially we shall change $u^*(t)$ to the value u_1 near $t = t_1$, and keep $u^*(t)$ unchanged otherwise. More exactly

$$u(t, \varepsilon, t_1, \ell_1, u_1) = \begin{cases} u_1 & \text{on} \quad t_1 - \ell_1 \varepsilon \leq t \leq t_1 \\ u^*(t) & \text{elsewhere on} \quad 0 \leq t \leq 1. \end{cases}$$

We abbreviate this controller by $u_\pi(t, \varepsilon)$ and the corresponding periodic response $x^*_\pi(t, \varepsilon)$.

The significance of this type of perturbation is that, at the time t_1, the solution initiating at x^*_0 is jerked or displaced (to first order in ε) by the vector

$$v(t_1) = [f(x^*(t_1), u_1) - f(x^*(t_1), u^*(t_1))]\, \ell_1 .$$

If this vector is then transported along the flow of the system $\dot{x} = f(x, u^*(t))$, we have the final displacement $v(1) = A_{1t_1} v(t_1)$, where A_{tt_1} is the fundamental solution matrix of the variational system

$$\dot{v} = \frac{\partial f}{\partial x} (x^*(t), u^*(t), \quad A_{t_1 t_1} = I.$$

It is the displacement vector $v(1)$ that enters onto the standard formulation of the maximal principle, where the initial point x_0^* is fixed. But in our situation of limit cycle control a more intricate analysis is required.

In order to locate the periodic response to $u_\pi(t, \varepsilon)$ we shall compute the initial point $x_0(\varepsilon, \pi)$ by the implicit function relation

$$x_\pi(1, x_0, \varepsilon) - x_0 = 0 .$$

Here $x_\pi(t, x_0, \varepsilon) = x(t, x_0, \varepsilon t_1, \ell_1, u_1)$ is the solution of $\dot{x} = f(x, u_\pi(t, \varepsilon))$ initiating at the point x_0. Upon solving this implicit relation for $x_0(\varepsilon, \pi)$, with $x_0(0, \pi) = x_0^*$, we obtain the periodic solution

$$x_\pi^*(t, \varepsilon) = x_\pi(t, x_0(\varepsilon, \pi), \varepsilon) \quad .$$

By definition $x_0(\varepsilon, \pi) = x_\pi^*(1, \varepsilon)$ so we can compute the perturbation in the amplitude using either $\frac{\partial x_0}{\partial \varepsilon}(0, \pi)$ or equally well $\frac{\partial x_\pi^*}{\partial \varepsilon}(1, 0)$. If we differentiate the implicit relation for $x_0(\varepsilon, \pi)$ we obtain at $\varepsilon = 0$,

$$\frac{\partial x}{\partial x_0}(1, x_0^*) \frac{\partial x_0}{\partial \varepsilon}(0, \pi) + \frac{\partial x_\pi}{\partial \varepsilon}(1, x_0^*, 0) - \frac{\partial x_0}{\partial \varepsilon}(0, \pi) = 0,$$

or

$$\frac{\partial x_0}{\partial \varepsilon}(0, \pi) = (I - P)^{-1} \frac{\partial x_\pi}{\partial \varepsilon}(1, x_0^*, 0).$$

The standard discussion of the maximal principle notes that

$$v(1) = \frac{\partial x_\pi}{\partial \varepsilon}(1, x_0^*, 0) = A_{1t_1} [f(x^*(t_1), u_1) - f(x^*(t_1), u^*(t_1))] \ \ell_1 .$$

Thus our present form of the maximal principle is

$$-(1,0,0,\ldots,0)\,(I-P)^{-1}\,A_{1t_1}\,[f(x^*(t_1),u_1) - f(x^*(t_1),u^*(t_1))] \geq 0.$$

If we transport the vector $\eta(1) = (1,0,0,\ldots,0)\,(I-P)^{-1}$ back along the flow according to the adjoint variational equation

$$\dot{\eta} = -\eta\,\frac{\partial f}{\partial x}\,(x^*(t),u^*(t)) \quad ,$$

then we can conclude that (since $\eta(t)\,v(t) =$ constant)

$$-\eta^*(t_1)\,[f(x^*(t_1),u_1) - f(x^*(t_1),u^*(t_1)] \geq 0$$

or

$$\eta^*(t_1)\,f(x^*(t_1),u^*(t_1)) \geq \eta^*(t_1)\,f(x^*(t_1),u_1)$$

at each instant t_1 and for each value $u_1 \in \Omega$. This is the required formulation of the maximal principle for our problem.

Finally let us comment on some of the technical difficulties encountered in completing the details of the above proof sketch.

It is easy to see that $x_\pi(t,x_0,\varepsilon) = x(t,x_0,\varepsilon,t_1,\ell_1,u_1)$ is continuous jointly in all arguments, since

$$\int_0^1 |\,u(t,\varepsilon_1,t_1,\ell_1,u_1) - u(t,\varepsilon_2,t_2,\ell_2,u_2)\,|\,dt \to 0$$

as $(\varepsilon_1,t_1,\ell_1,u_1) \to (\varepsilon_2,t_2,\ell_2,u_2)$. It is slightly harder to verify that

$\frac{\partial x}{\partial x_0}(t,x_0,\varepsilon,t_1,\ell_1,u_1)$ is also continuous in $(t,x_0,\varepsilon,t_1,\ell_1,u_1)$. However this last assertion follows from the observation that $Z(t) = \frac{\partial x}{\partial x_0}(t,x_0,\varepsilon,t_1,\ell_1,u_1)$ is the fundamental solution matrix of

$$\dot{Z} = \frac{\partial f}{\partial x}\,(x(t,x_0,\varepsilon,t_1,\ell_1,u_1),\ u(t,\varepsilon,t_1,\ell_1,u_1))\ Z$$

with $Z(0) = I$. Incidentally this calculation validates our use of the implicit function theorem to define $x_0(\varepsilon, t_1, \ell_1, u_1) = x_0(\varepsilon, \pi)$ which is continuous in all arguments jointly. A more refined study shows that $\frac{\partial x_\pi}{\partial \varepsilon}(1, x_0, 0)$ exists, and so $\frac{\partial x_0}{\partial \varepsilon}(0, \pi)$ is just as computed above.

As in the standard development of the maximal principle we need to take t_1 as a Lebesgue time for $u^*(t)$ and $f(x^*(t), u^*(t))$ so that the perturbation

$$\frac{\partial x_\pi}{\partial \varepsilon}(1, x_i^*, 0) = A_{1t_1}[f(x^*(t_1), u_1) - f(x^*(t_1), u^*(t_1))]\,\ell_1$$

is correct. Also the use of $\frac{\partial x_\pi^*}{\partial \varepsilon}(1, 0)$ to approximate $x_\pi^*(1, \varepsilon) - x^*(1)$ must be justified as in the standard maximal principle. We omit any further study of the details of the proof.

REFERENCES

1. E. B. Lee and L. Markus, Foundations of Optimal Control Theory, Wiley 1967.

2. D. Spyker, Optimal Control of Cardiac Assist Devices, Ph.D. Thesis, Univ. of Minnesota 1969.

The Stable Manifold Theorem

Via an Isolating Block

Richard McGehee

1. Introduction

Let f be a diffeomorphism of a smooth manifold. We say a fixed point x_0 is hyperbolic if $Df(x_0)$ has no eigenvalues of modulus 1. Given a neighborhood U of x_0 we define the local positively asymptotic set:

$$W^+ = \{ x \in U \colon f^k(x) \in U \text{ for } k > 0, \text{ and } f^k(x) \to x_0 \text{ as } k \to \infty \}$$

The local stable manifold theorem states that, for small U, W^+ is an embedded submanifold of U, with the embedding as smooth as f.

The local stable manifold theorem has a long history dating back to Poincaré. (See Hartman's notes [2, p. 271].) The standard proof uses power series techniques in the analytic case and the contraction mapping priniciple in the C^r case. More modern proofs use the implicit function theorem in a Banach space. In this paper we exploit the concept of an isolating block as defined by Conley and Easton [1] to give a proof using only elementary topology of Euclidean spaces and elementary linear algebra. Techniques similar to those presented in this paper have been used in certain case of the three-body problem to prove that the set of parabolic orbits is a smooth submanifold [3].

In the next section we give a precise statement of the local stable manifold theorem in Euclidean space. In section 3 we develop properties of an isolating block which we use in section 4 to prove the theorem in the Lipschitz case. In section 5 we complete the proof of the theorem.

This research was supported by NSF Grant GP 27275

2. Preliminaries

We shall state the local stable manifold theorem in Euclidean space, but first we must introduce some notation.

Fix integers m and n and fix norms (not necessarily Euclidean) on R^m and R^n. We denote both these norms by $\| \ \|$. For $(x, y) \in R^m \times R^n$, let

$$\|(x, y)\| = \max \ (\|x\|, \|y\|).$$

We further use $\| \ \|$ to denote the linear operator norm subordinate to $\| \ \|$. We use these norms to define the unit discs:

$$I_1 = \{x \in R^m : \|x\| \leq 1\}$$

$$I_2 = \{y \in R^n : \|y\| \leq 1\}$$

$$I = \{(x, y) \in R^m \times R^n : \|(x, y)\| \leq 1\} = I_1 \times I_2$$

Let $A \in GL(R^m \times R^n)$, the set of linear isomorphisms on $R^m \times R^n$. We shall say A is <u>canonically hyperbolic</u> if

$$A = \begin{bmatrix} A_1 & 0 \\ 0 & A_2 \end{bmatrix},$$

where $A_1 \in GL(R^m)$, with $\|A_1\| < 1$, and $A_2 \in GL(R^n)$, with $\|A_2^{-1}\| < 1$.

Let $p_\epsilon : R^m \times R^n \to R^m \times R^n$ for small real ϵ. We shall say p_ϵ is C^r-<u>small</u> if $D^s p_\epsilon$ (the sth derivative of p_ϵ) is continuous and approaches zero as $\epsilon \to 0$ for $s = 0, 1, \ldots, r$. We shall say p_ϵ is Lip^r-<u>small</u> if p_ϵ is C^r-small and $D^r p_\epsilon$ is Lipschitz, with the Lipschitz constant approaching zero as $\epsilon \to 0$. We shall say p_ϵ is <u>smooth</u> if p_ϵ is either C^r-small, $r \geq 1$, or Lip^r-small, $r \geq 0$.

Let $f : R^m \times R^n \to R^m \times R^n$. Define the local stable and unstable manifolds as:

$$W^+(f) = \{ z \in I: \ f^k(z) \in I \quad \text{for all} \ k > 0 \},$$

$$W^-(f) = \{ z \in I: \ f^k(z) \in I \quad \text{for all} \ k < 0 \} .$$

Note that, if A is canonically hyperbolic, $W^+(A) = I_1 \times \{0\}$ and $W^-(A) = \{0\} \times I_2$.

By suitably choosing a coordinate patch we can reduce the local stable manifold theorem as stated in the introduction to the following:

<u>Theorem 1:</u> <u>Let</u> A <u>be canonically hyperbolic, let</u> p_ϵ <u>be smooth, and let</u> $f = A + p_\epsilon$. <u>Then, for small</u> ϵ, $W^+(f)$ <u>is the graph of a function</u> $\varphi : I_1 \to I_2$. <u>Furthermore,</u> φ <u>is as smooth as</u> p_ϵ . I. e., <u>if</u> p_ϵ <u>is</u> Lip^r-small, $r \geq 0$, <u>then</u> φ <u>is</u> Lip^r; <u>if</u> p_ϵ <u>is</u> C^r-small, $r \geq 1$, <u>then</u> φ <u>is</u> C^r.

By considering f^{-1}, one also concludes that $W^-(f)$ is a smooth submanifold. W^+ and W^- intersect at exactly one point, a fixed point for f. Thus W^+ and W^- are the stable and unstable manifolds to that fixed point.

3. The Isolating Block

The unit disc $I \subset R^m \times R^n$ is an isolating block for f in the sense defined by Conley and Easton [1]. For our purposes, the important properties are those listed in the proposition below. Let π_1 and π_2 be the projection maps on $R^m \times R^n$: $\pi_1(x, y) = x$, $\pi_2(x, y) = y$. Also let

$$\mathscr{V} = \{ (x, y) \in R^m \times R^n : \|y\| > \|x\| \}.$$

<u>Proposition</u> 2. <u>Let</u> A <u>be canonically hyperbolic, let</u> p_ϵ <u>be</u> C^0- small, <u>and let</u> $f = A + p_\epsilon$. <u>Then for small</u> ϵ,

(3-1) $\pi_1 f(I) \subset I_1$,

(3-2) $f: I_1 \times \partial I_2 \to I_1 \times (R^n - I_2)$ is a homotopy equivalence.

Furthermore, if p_ϵ is Lip^0-small, then there exists a $\nu > 0$ such that, if $z_1, z_2 \in I$, with $z_1 - z_2 \in \mathcal{V}$, then

(3-3) $f(z_1) - f(z_2) \in \mathcal{V}$

(3-4) $\| \pi_2(f(z_1) - f(z_2)) \| \geq \nu \| \pi_2(z_1 - z_2) \|$

Proof. Let $A = \begin{bmatrix} A_1 & 0 \\ 0 & A_2 \end{bmatrix}$, with $\|A_1\| = \mu_1 < 1$ and $\|A_2^{-1}\| = \mu_2^{-1} < 1$.

Since $\|A_1\| < 1$, (3-1) follows immediately.

Consider (3-2). Since ∂I_2 is a strong deformation retract of $R^n - A_2^{-1}(I_2)$, $A_2(\partial I_2)$ is a strong deformation retract of $R^n - I_2$. Therefore, since $A_1(I_1)$ is a strong deformation retract of I_1, $A(I_1 \times \partial I_2) = A_1(I_1) \times A_2(\partial I_2)$ is a strong deformation retract of $I_1 \times (R^n - I_2)$. Therefore $A: I_1 \times \partial I_2 \to I_1 \times (R^n - I_2)$ is a homotopy equivalence. For small ϵ, f and A restricted to $I_1 \times \partial I_2$ are homotopic by the homotopy

$$H: (I_1 \times \partial I_2) \times [0, \epsilon] \to I_1 \times (R^n - I_2)$$

$$H(z, t) = Az + P_t(z).$$

Thus $f: I_1 \times \partial I_2 \to I_1 \times (R^n - I_2)$ is a homotopy equivalence.

Now consider (3-3) and (3-4). For $z \in \mathcal{V}$, we have

$$\|\pi_2 A_z\| \geq \mu_2 \|\pi_2 z\|$$

$$\|\pi_1 A_z\| \leq \mu_1 \|\pi_1 z\| < \mu_1 \|\pi_2 z\|$$

Let $\delta > 0$ be such that $\mu_2 - \delta = \nu > 1$ and $\mu_2 - \mu_1 - 2\delta > 0$. Choose ε so small that the Lipschitz constant of p_ε is less that δ. Then, for $z_1 - z_2 \in \mathcal{V}$,

$$\|\pi_2(f(z_1) - f(z_2))\| \geq \|\pi_2 A(z_1 - z_2)\| - \|\pi_2(p_\varepsilon(z_1) - p_\varepsilon(z_2))\|$$

$$\geq (\mu_2 - \delta)\|\pi_2(z_1 - z_2)\|,$$

and we have established (3-4). Also,

$$\|\pi_1(f(z_1) - f(z_2))\| \leq \|\pi_1 A(z_1 - z_2)\| + \|\pi_1(p_\varepsilon(z_1) - p_\varepsilon(z_2))\|$$

$$\leq (\mu_1 + \delta)\|\pi_2(z_1 - z_2)\|,$$

and hence,

$$\|\pi_2(f(z_1) - f(z_2))\| - \|\pi_1(f(z_1) - f(z_2))\| \geq (\mu_2 - \mu_1 - 2\delta)\|\pi_2(z_1 - z_2)\|.$$

Thus we have established (3-3) and completed the proof of Proposition 2.

4. The Lipschitz Case

In this section we prove Theorem 1 when p_ε is Lip^0-small. We begin with some more notation.

Define the set of vertical discs:

$$V = \{\Gamma \subset I : \pi_2 \Gamma = I_2, \text{ and } z_1, z_2 \in \Gamma \Rightarrow z_1 - z_2 \in \mathcal{V} \cup \{0\}\}.$$

Thus a vertical disc Γ is the graph of a Lipschitz function $I_2 \to I_1$ and hence is an embedded n-disc. Furthermore, the inclusion map $\partial\Gamma \subset I_1 \times \partial I_2$ is a homotopy equivalence.

Proposition 3. $\Gamma \in V \Rightarrow f(\Gamma) \cap I \in V$.

Proof. By (3-3), $f(z_1)$ and $f(z_2) \in \Gamma \Rightarrow f(z_1)-f(z_2) \in \mathcal{V} \cup \{0\}$, so we need only show that $L_2 \subset \pi_2(f(\Gamma) \cap I)$. Suppose there is a $y_0 \in L_2$ with $y_0 \notin \pi_2 f(\Gamma)$. Then by (3-2) the commutative diagram

$$\begin{array}{ccc} \partial\Gamma & \subset & \Gamma \\ \big\downarrow \pi_2 f & & \big\downarrow \pi_2 f \\ R^n - L_2 & \subset & R^n - \{y_0\} \end{array}$$

is a homotopy equivalence of S^{n-1} factoring through an n-disc. Since this is a contradiction, we must have $L_2 \subset \pi_2 f(\Gamma)$. By (3-1), $L_2 \subset \pi_2(f(\Gamma) \cap I)$ and the proposition is proved.

We can now prove Theorem 1 when p_ϵ is Lip^0-small. That is, we can now prove the following lemma:

Lemma 4. Let A be canonically hyperbolic, let p_ϵ be Lip^0-small, and let $f = A + p_\epsilon$. Then $W^+(f)$ is the graph of a Lipschitz function $\varphi: L_1 \to L_2$.

Proof. It is sufficient to show that any vertical disc intersects W^+ at exactly one point. Thus the proof of Lemma 4 will be complete when we have established the following proposition:

Proposition 5. If $\Gamma \in V$, then $\Gamma \cap W^+$ contains exactly one point.

Proof. Let $\Gamma_0 = \Gamma$, let $\Gamma_k' = f(\Gamma_{k-1}') \cap I$ for $k = 1, 2, \ldots$, and let $\Gamma_k = f^{-k}(\Gamma_k')$. Then $\{\Gamma_k\}$ is a nested sequence of non-empty discs, hence $\cap \Gamma_k \neq \phi$. Also, if z_1 and $z_2 \in \cap \Gamma_k$, then by (3-4),

$$1 \geq \| \pi_2(f^k(z_1) - f^k(z_2)) \| > \nu^k \|\pi_2(z_1 - z_2)\|$$

for all k. Thus $z_1 = z_2$ and $\cap \Gamma_k$ contains exactly one point. Since $\cap \Gamma_k = \Gamma \cap W^+$, the proposition is proved.

5. The Differentiably Cases

Next we consider Theorem 1 in the case when p_ϵ is C^1-small. By Lemma 4 we know there is a Lipschitz function φ whose graph is the stable manifold. We must find a function $\psi: I_1 \to L(R^m, R^n)$, the space of linear maps from R^m to R^n, such that $\psi = D\varphi$. Our plan is to define a new map on $R^m \times L(R^m, R^n)$ and use the techniques of the previous section to construct a stable manifold for the new map. This stable manifold will be the graph of ψ.

To begin we introduce some more notation. For $B \in GL(R^m \times R^n)$ and $\xi \in L(R^m, R^n)$, we define $B*\xi \in L(R^m, R^n)$ by

$$B*\xi = (\pi_2 B(id, \xi))(\pi_1 B(id, \xi))^{-1} .$$

Note that $A*\xi = A_2 \xi A_1^{-1}$ for canonically hyperbolic A. Thus A* is a linear map on $L(R^m, R^n)$ with $\|(A*)^{-1}\| \leq \|A_1\| \; \|A_2^{-1}\| < 1$. Denote by I_3 the unit disc in $L(R^m, R^n)$. For B near A, B* defines a map

$$B* : \quad I_3 \to L(R^m, R^n)$$

with the property

$$(5\text{-}1) \qquad B_1^* \circ B_2^* = (B_1 \circ B_2)^*.$$

Let $\mathcal{A}$ be a neighborhood of A such that B* is defined on I_3 for $B \in \mathcal{A}$. Note that the map

$$\mathcal{A} \times I_3 \to L(R^m, R^n)$$

$$(B, \xi) \to B*\xi$$

is C^{∞}. Since $\|(A^*)^{-1}\| < 1$, there exist a $\vartheta > 1$ and a neighborhood $\mathcal{a}'$ of A such that $B \in \mathcal{a}' \Rightarrow \|(DB^*(\xi))^{-1}\| \leq \vartheta^{-1}$ for $\xi \in I_3$. Thus for $\xi_1, \xi_2 \in I_3$ and $B \in \mathcal{a}'$,

(5-2) $$\|B^*\xi_1 - B^*\xi_2\| \geq \vartheta \|\xi_1 - \xi_2\| .$$

Now let $f = A + p_\epsilon$, with A canonically hyperbolic and p_ϵ C^1-small. By Lemma 4 there is a function φ whose graph is the stable manifold $W^+(f)$. Let $\underset{\sim}{I} = I_1 \times I_3$ and define

$$\underset{\sim}{f} : \underset{\sim}{I} \to I_1 \times L(R^m, R^n)$$

$$\underset{\sim}{f}(x, \xi) = (\pi_1 f(x, \varphi(x)), \ (Df(x, \varphi(x)))^* \xi).$$

Then $\underset{\sim}{f} = \underset{\sim}{A} + \underset{\sim}{p}_\epsilon$, where $\underset{\sim}{A} = \begin{bmatrix} A_1 & 0 \\ 0 & A^* \end{bmatrix}$ and $\underset{\sim}{p}_\epsilon$ is C^0-small.

Furthermore, if p_ϵ is $C^r(\mathrm{Lip}^r)$-small, then $\underset{\sim}{p}_\epsilon$ is $C^{r-1}(\mathrm{Lip}^{r-1})$-small.

Let

$$\underset{\sim}{W}^+ = \{ \underset{\sim}{z} \in \underset{\sim}{I} : \ \underset{\sim}{f}^k(\underset{\sim}{z}) \in \underset{\sim}{I}, \ k > 0 \}.$$

If p_ϵ is Lip^1-small, Lemma 4 can be applied to the map $\underset{\sim}{f}$ to conclude that $\underset{\sim}{W}^+$ is the graph of a Lipschitz function $\psi : I_1 \to I_3$. However, a stronger result is true:

<u>Proposition</u> 6. <u>If</u> p_ϵ <u>is</u> $C^1(\mathrm{Lip}^1)$-small, <u>then</u> $\underset{\sim}{W}^+$ <u>is the graph of a continuous (Lipschitz) function</u> $\psi : I_1 \to I_3$.

<u>Proof</u>. Let $\underset{\sim}{V} = \{ \{x_0\} \times I_3 : \ x_0 \in I_1 \}$. By (3-1) and (3-2) of Proposition 2, $\pi_1 \underset{\sim}{f}(\underset{\sim}{I}) \subset I_1$ and $\underset{\sim}{f} : I_1 \times \partial I_3 \to I_1 \times (L(R^m, R^n) - I_3)$ is a homotopy

equivalence. Therefore, the argument used in the proof of Proposition 3 gives us

$$\Gamma \in \underset{\sim}{V} \Rightarrow \underset{\sim}{f}(\Gamma) \cap \underset{\sim}{I} \in \underset{\sim}{V}.$$

Now choose ϵ so small that $Df(z) \in \mathcal{a}'$ for $z \in I$. Then by (5-2) we have

$$\|Df(z))^* \xi_1 - (Df(z))^* \xi_2 \| \geq \nu \| \xi_1 - \xi_2 \| \tag{5-3}$$

for $z \in I$ and $\xi_1, \xi_2 \in I_3$. Thus the arguments used in the proof of Proposition 4 can be applied to $\underset{\sim}{f}$ and $\underset{\sim}{V}$ to conclude that $\Gamma \cap \underset{\sim}{W}^+$ contains exactly one point for any $\Gamma \in \underset{\sim}{V}$. Therefore $\underset{\sim}{W}^+$ is the graph of a function $\psi: I_1 \to I_3$. Note that $\underset{\sim}{W}^+ = \cap \underset{\sim}{f}^{-k}(\underset{\sim}{I}) \cap \underset{\sim}{I}$ and is therefore compact. Thus π_1 restricted to $\underset{\sim}{W}^+$ is a homeomorphism and hence ψ is continuous. The proof of Proposition 6 is complete.

If we can now show that $\psi = D\varphi$, then we have completed the proof of Theorem 1 when p_ϵ is C^1 or Lip^1-small. Thus we have only left to prove the following:

Proposition 7. Let $(x_0, \xi_0) \in \underset{\sim}{W}^+$. Then $\xi_0 = D\varphi(x_0)$.

Proof. For $\xi: R^m \to R^n$, let

$$\text{graph}(\xi) = \{(x, y) \in R^m \times R^n : y = \xi(x)\}.$$

For $\mathcal{U} \in L(R^m, R^n)$, let

$$\text{graph}(\mathcal{U}) = \cup \{\text{graph}(\xi) \in \mathcal{U}\}.$$

In section 4 we proved that φ is Lipschitz, i.e.

$$W^+ = \text{graph}(\varphi) \subset (x, \varphi)) + \text{graph}(I_3), \tag{5-4}$$

where " + " indicates vector space addition.

Now let $z_0 = (x_0, \varphi(x_0))$. It is sufficient to show that, given any neighborhood $\mathcal{U}$ of ξ_0, there exists a neighborhood U of z_0 such that

$$W^+ \cap U \subset z_0 + \text{graph}(\mathcal{U}).$$

By (5-1) and (5-3), there exists a positive integer k such that $I_3 \subset (Df^k(z_0))^*(\mathcal{U})$. Thus there exists a neighborhood U' of $f^k(z_0)$ such that

$$f^{-k}(U' \cap (f^k(z_0) + \text{graph}(I_3))) \subset z_0 + \text{graph}(\mathcal{U}).$$

Let $U = f^k(U')$. By (5-4), $W^+ \subset f^k(z_0) + \text{graph}(I_3)$, and hence

$$f^{-k}(W^+ \cap U') = W^+ \cap U \subset z_0 + \text{graph}(\mathcal{U}).$$

Hence $\xi_0 = D\varphi(x_0)$ and the proof of Proposition 7 is complete.

We can now prove Theorem 1 by induction r.

<u>Proof of Theorem 1.</u> Lemma 4 establishes the theorem when p_ϵ is Lip^0-small. Propositions 6 and 7 establish the theorem when p_ϵ is Lip^1 or C^1-small. We proceed by induction.

Suppose p_ϵ is Lip^r or C^r-small, $r \geq 2$. Then $\underset{\sim}{p}_\epsilon$ is Lip^{r-1} or C^{r-1}-small. By inductive hypothesis, $\underset{\sim}{W}^+$ is the graph of a Lip^{r-1} or C^{r-1} function ψ. By Proposition 7, $\psi = D\varphi$. Hence φ is Lip^r or C^r and the proof of Theorem 1 is complete.

References

1. C. Conley and R. Easton, "Isolated Invariant Sets and Isolating Blocks," <u>Trans. Amer. Math. Soc.</u>, Vol. 158, No.1 (1971), 35-61.

2. P. Hartman, <u>Ordinary Differential Equations</u>, John Wiley and Sons, New York, 1964.

3. R. McGehee, "A Stable Manifold Theorem for Degenerate Fixed Points with Applications to Celestial Mechanics," (to appear).

Stability of a Lurie Type Equation

K. R. Meyer

In their study of nonlinear electrical circuits Brayton and Moser [1] investigated the asymptotic behavior of a system of nonlinear differential equations that describe the state of an electrical network. The aim was to give conditions that insure nonoscillating solutions. The criterion obtained in [1] was very restrictive and Moser in [2] obtained more general criteria by using the method of Popov of automatic control theory. The method of Popov has been very successful in the study of the stability properties of the Lurie equations (see [3] for a detailed discussion).

At first glance the equations of Brayton and Moser bear no resemblance to the usual Lurie equations but this note will show that by a change of variables the equations take a form similar to the Lurie equations. Once the equations are written in this new form it is then clear how to use the methods developed in control theory to study their stability properties. In particular it is clear that Popov's method would yield a stability criterion. It is also clear how to construct a Liapunov function for these equations. We choose the latter to reprove Moser's theorem in a straightforward way.

The system considered in [1, 2] is of the form

$$\begin{aligned} \dot{x} &= -Ax + By \\ \dot{y} &= Cx - f(y) \end{aligned} \tag{1}$$

This research was supported by ONR contract number N00014-67-A-0113-0019

where x is an n-vector, y an m vector, A, B and C are constant matrices of appropriate dimensions, A nonsingular, and f is an m vector valued function of the m vector y.

One wishes to find conditions on A, B, C and f so that all solutions of (1) approach a finite number of equilibrium states and hence rule out oscillatory behavior. The fundamental assumption on f is that it can be written in the form $f(y) = \nabla G(y) - cA^{-1}By$ where G is a scalar function and ∇ stands for gradient. It is also assumed that G tends to infinity as y tends to ∞ and G has a finite number of critical points $(y_1, \ldots, y_k)$.

Moser then obtains conditions on the coefficients A, B and C such that all solutions of (1) tend to $x = 0$, $y = y_j$, $j = 1, \ldots, k$.

If one makes the change of variables $u=x$, $v=-y -cA^{-1}x$ and lets $K =-(A+BCA^{-1})$, $D=-CA^{-1}$ then the Equations (1) become

$$
\begin{aligned}
&\dot{u} = Ku - Bv \\
(2)\qquad &\dot{v} = \nabla G(y) \\
&y = Du - v
\end{aligned}
$$

If y is a scalar and $G(y) = \int_0^y \varphi(\tau)\, d\tau$ then Equations (2) reduce to the usual indirect control equations of Lurie.

Thus it is natural to make the Lefschetz change of variables $z = Ku - Bv$, $\sigma = Du - v$ which is nonsingular provided

$$
(3)\qquad \begin{vmatrix} K & -B \\ D & -I \end{vmatrix} = |K|\,|-I - DK^{-1}B| \neq 0 \quad .
$$

This condition is clearly necessary for isolated equilbrium points. Under this change of variables the Equations (2) become

$$
\begin{aligned}
&\dot{z} = Kz - B\nabla G(\sigma) \\
(4)\qquad &\dot{\sigma} = Dz - \nabla G(\sigma)
\end{aligned}
$$

For Equations (4) the natural Liapunov function is of the form

(5) $$V = z'Pz + G(\sigma)$$

where P is a positive definite symmetric matrix

We shall show that one can give conditions on the coefficients of (4) such that one can find a P that makes V a Liapunov function for (4). The existence of such a P is the result of a lemma by Anderson [4]. Anderson's lemma is a natural generalization of the Kalman-Yarnkovich lemma discussed in [3]. Henceforth we shall assume that (K, B, D,) is a completely controllable, completely observable triple. This assumption is necessary for the application of the lemma of Anderson but one could dispense with this assumption by using the methods developed in [5].

The condition on the coefficients of (4) are stated in terms of the transfer function

(6) $$T(\lambda) = I + D\{\lambda I - K\}^{-1} B.$$

An $m \times m$ matrix function Z of a complex variable λ is called <u>positive real</u> if

i) the elements of Z are rational functions with no poles for $\mathrm{Re}\lambda > 0$

ii) $\overline{Z(\lambda)} = Z(\overline{\lambda})$

iii) $Z(\overline{\lambda})^T + Z(\lambda)$ is nonnegative definite for $\mathrm{Re}\lambda > 0$. Z is called <u>strictly positive real</u> if i) holds for $\mathrm{Re}\lambda \geq 0$, ii) holds and $Z(\overline{\lambda})^T + Z(\lambda)$ is positive definite for $\mathrm{Re}\lambda \geq 0$.

The main theorem is then

<u>Theorem 1</u>. <u>If</u> $T(\lambda) = I + D\{\lambda I - K\}^{-1} B$ <u>is positive real then all solutions of</u> (4) <u>are bounded and if it is strictly positive real all solutions of</u> (4) <u>approach one of the equilibrium points</u> $(0, \sigma_i)$ <u>where</u> σ_i <u>is such that</u> $\nabla G(\sigma_i) = 0$.

We can state Theorem 1 for the original system of Equations (1) by tracing back the coordinate changes. In terms of the original matrices

(7)
$$
\begin{aligned}
T(\lambda) &= I + (-CA^{-1})(\lambda I + A + BCA^{-1})^{-1}B \\
&= I - C\{\lambda A + A^2 + BC\}^{-1}B \\
&= I - C\{\lambda A + A^2\}^{-1}B\,\{I + C(\lambda A + A^2)^{-1}B\}^{-1} \\
&= \{I + C(\lambda A + A^2)^{-1}B\}^{-1}
\end{aligned}
$$

thus

Corollary 1. *If* $T(\lambda)^{-1} = \{I + C(\lambda A + A^2)^{-1}B\}$ *is strictly positive real and* (3) *holds them all solutions of* (1) *approach one of the equilibrium points* $(0, y_i)$ *where* y_i *is a critical points of* G.

Remark. Moser does not assume that (3) hold explicitly but one can easily show that (3) is equivalent to the condition that the residue at ∞ of $T(\lambda) + T(\bar{\lambda})^*$ is nonsingular. This is an easy consequence of Moser's condition.

Proof of Theorem 1.

We prove Theorem 1 by using the lemma given below to show that there exists a Liapunov function of the form (5) for (4).

Anderson's Lemma: *If* $T(\lambda) = I + D(\lambda I + K)^{-1}B$ *is positive real then there exists a positive definite* $n \times n$ *matrix* P *and an* $m \times n$ *matrix* L *such that*

(8)
$$
\begin{aligned}
PK + K'P &= -LL' \\
PB &= L - \tfrac{1}{2}D \quad .
\end{aligned}
$$

In the proof of this lemma one has also the following matrix identity

(9)
$$
(m^*(i\omega)L - I)(L'm(i\omega) - I) = I + \frac{1}{2}\{C'm(i\omega) + m^*(i\omega)C\}
$$

where $m(i\omega) = (i\omega I - K)^{-1}B$ and $*$ denotes conjugate transpose. One sees at once that the right hand side of (8) is $\frac{1}{2}\ \{T(i\omega) + T'(-i\omega)\}$ and so if T is strictly positive real then

(10) $$I - L'(i\omega I - K)^{-1}B = (I - L'm(i\omega))$$

is nonsingular for all real ω. This fact is useful in the analysis of set where $\dot{V}$ is identically zero.

Let the P in (5) be as given by Anderson's lemma then the derivative $\dot{V}$ of V along the trajectories of (4) is given by

$$-\dot{V} = -z^T\{K'P + PK\}\ z + 2z'\ \{PB - \tfrac{1}{2}D'\}\ \nabla G + \nabla G'\nabla G = \|\nabla G + L'z\|^2$$

Since $V \to \infty$ as z and $\sigma \to \infty$ and $\dot{V} \leq 0$ it follows by the well-known Liapunov theorems that all solutions of (4) are bounded.

In order to conclude that all solutions of (4) tend to $(0, \sigma_0)$ we must use the theorem of LaSalle [6] that states that all solutions tend to the largest invariant set of (4) that is contained in the set where $\dot{V} \equiv 0$. Thus Theorem 1 is established once one shows that the largest invariant set contained in the set where $\dot{V} \equiv 0$ is the set $\{(0, \sigma_1), \dots, (0, \sigma_k)\}$.

Let $z(t)$, $\sigma(t)$ be a solution of (4) that is such that $z(t) \not\equiv 0$ and $\dot{V}(z(t), \sigma(t)) \equiv 0$. Then $G(\sigma(t)) = -Lz(t)$ and so $z(t)$ satisfies

$$\dot{z} = \{K + BL'\}\ z$$

But $\dot{z}$ is bounded for all t and so the matrix $K + BL'$ must have an eigenvalue on the imaginary axis. The characteristic equation for $K + BL'$ is

$$|\lambda I - K - BL'| = |\lambda I - K|\ |I - (\lambda I - K)^{-1}BL'| = |\lambda I - K|\ |I - L'(\lambda I - K)^{-1}B|$$

But we have seen that if $T(\lambda)$ is strictly positive real the matrix $I - L'(\lambda I - K)^{-1}B$ is nonsingular for $\lambda = i\omega$, ω real. Hence $z(t) \equiv 0$, $\sigma(t) \equiv 0$.

References

1. R. K. Brayton and J. K. Moser, "A theory of nonlinear networks I," Quart, of Appl. Math., 22 (1964) pp. 1-33.

2. J. K. Moser, "On nonoscillating networks," Quart. of Appl. Math., 25 (1967) pp. 1-9.

3. S. Lefschetz, Stability of nonlinear control systems, Academic Press, New York, 1965.

4. B. D. O. Anderson, "A system theory criterion for positive real matrices," SIAM Journal on Control, 5 (1967) pp. 171-182.

5. K. R. Meyer, "On the existence of Lyapunov functions for the problem of Lurie," J. SIAM Control, Ser. A, 3 (1966) pp. 373-383.

6. J. P. LaSalle, "The extent of asymptotic stability," Proc. Math. Acad. Sci., 46 (1954) pp. 363-365.

A Nonlinear Integral Equation Relating Distillation Processes

E. R. Rang

1. Introduction

A functional equation that relates a true boiling point distillation with a differential distillation, as defined by Rayleigh's Equation, is derived. It is a nonlinear integral equation which can be solved by numerical computation. The calculation is applied to data on gasoline and hydrocarbon mixtures.

2. Distillation Processes

Commerical distillations are usually carried out in a continuous manner. The conditions are set so that as the material enters the column a portion of it is vaporized. It is assumed for design that the vapor is in equilibrium with the remaining liquid so the process is called an equilibrium flash vaporization. Our discussion does not include such systems but rather is concerned with laboratory distillations of batches of material, in particular, gasoline or other mixtures of hydrocarbons.

The results of a procedure, usually displayed as plots of the current temperature against the fraction of the batch that has been distilled, change significantly with distillation rate and the amount of reflux that occurs. In this context, reflux is an unquantified term which refers to the process of repeated condensations and vaporizations in the upper raches of the distillation flask or in the column. It enhances the separation of components of the mixture. An ideal process that completely fractionates the mixture is called a true boiling point distillation. It may be considered as a process with infinite reflux. With time and patience some laboratory procedures approximate this. The ASTM method does not aim for fractionation but provides an economical standard assay for comparing samples. In a nonfractionation distillation, the cut of material being distilled does not necessarily reflect its isolated boiling characteristics. However, the ASTM and the true boiling distillations are related and it is of practical and theoretical interest to have methods of estimating one from the other. This study is pointed in that direction although we idealize the ASTM procedure as a differential distillation. That is, one in which no reflux occurs and the vapor which is delivered to the condenser is assumed to be in equilibrium with the liquid remaining in the pot.

3. Correlations of Distillation Curves

Basic correlations between distillation curves were given by Geddes [1], Edmister and Okamoto [2, 3] and Van Winkle [4]. These empirical relations are represented graphically and generally a particular point and several slopes of the curves are correlated. Surveys of the various techniques may be found in References [5, 6]. It appears that the developments given by Hoffman [7, 8, 9] are essentially an empirical fitting of curves of the form

$$\Psi = \lambda \int_{T_0}^{T} e^{-h(\tilde{T} - T_m)^2} \, d\tilde{T} \tag{3.1}$$

for constants λ, h, T_m. The quantity Ψ is the mole fraction of the batch distilled after a temperature T has been reached. The parameters of the curves are related by applying correlations between the ordinates and slopes at the 50%-point for TBP-and ASTM-Distillations. The procedure used by Gandbhir and Virk [10] exploits the same relations. They plot the data on probability paper. Thus, if Equation (3.1) fits the data, the curves become straight lines on the probability rulings. Relations for equilibrium flash vaporizations are given in References [11, 12, 13].

The present study began as an attempt to derive a linear functional relation of the form used by Baker and Rang [14] to empirically relate distillation curves. We have not yet achieved this goal. The functional equation which is derived subsequently is nonlinear.

4. Finite Number of Components

We begin with the usual derivation of Rayleigh's equation, Van Winkle [5] p. 170 for example, in order to introduce notation. Consider a batch distillation of M_0 moles of a mixture which has n components with mole fractions $X_1, X_2, \ldots, X_n$. At a certain state assume M moles have been distilled off and $\overline{M}$ moles remain. Let $x_1, \ldots, x_n$ be the mole fractions of the components in $\overline{M}$ and let $z_1, \ldots, z_n$ be the mole fractions of the components in M. Then we can write material balance equations

$$(4.1)\qquad \begin{cases} M + \overline{M} = M_0 \\ Mx_i + Mz_i = M_0 X_i, \end{cases}$$

or with $\Psi = M/Mo$, the latter equation is

$$(4.2)\qquad (1 - \Psi)\, x_i + z_i = X_i \,.$$

Now suppose an increment M is distilled and it has mole fractions of its components $y_1, \ldots, y_n$ and is in equilibrium with the remaining liquid. The ith component changes by an amount

$$(4.3)\qquad (\overline{M} - \Delta M)\,(x_i + \Delta x_i) - x_i \overline{M} \cong - x_i \Delta M + \overline{M}\, \Delta x_i$$

in the liquid and by an amount $y_i \Delta M$ in the vapor. Hence, $y_i \Delta M = x_i \Delta M - \overline{M} \Delta x_i$. Rearranging gives

$$(4.4)\qquad \frac{\Delta x_i}{\Delta M} = \frac{x_i - y_i}{M}$$

and allowing a limit as $M = Mo\Delta\Psi \to 0$ yields Rayleigh's equation

$$(4.5) \qquad \frac{dx_i}{d\Psi} = \frac{x_i - y_i}{1 - \Psi} \ .$$

At $\Psi = 0$, $x_i = X_i$. The equation may be integrated in the form

$$(4.6) \qquad x_i = \frac{1}{1-\Psi}\left[X_i - \int_0^{\Psi} y_i \, d\Psi\right] \ .$$

Using Equation (4.2) gives

$$(4.7) \qquad z_i = \frac{1}{\Psi}\int_0^{\Psi} y_i \, d\Psi \ .$$

Along with the initial distribution X_i we assume we have the boiling points T_i of each of the components in a pure state. Then with

$$(4.8) \qquad \phi_i = \sum_{j=1}^{i} X_j, \ (\phi_0 = 0),$$

a plot of the step function

$$(4.9) \qquad T = T_i \ , \ \phi_{i-1} \leqq \phi < \phi_i \ , \ (T_i < T_{i+1}),$$

will be called the true boiling point curve of the mixture. Now we come to a fundamental hypothesis. Assume that there is a function $T(\Psi)$ and an equilibrium ratio $K_i(T, T_i)$ so that

$$(4.10) \qquad y_i = K(T(\Psi), \ T_i) \, x_i \ .$$

Then Rayleigh's Equation may be integrated to give

$$(4.11) \qquad x_i = \frac{X_i}{1-\Psi} \exp\left\{\int_0^{\Psi} \frac{-K_i(T(\alpha), T_i)}{1-\alpha} \, d\alpha\right\} \ .$$

Since the sum of the mole fraction is one, a functional equation characterizing the distillation curve $T(\Psi)$ is

$$(4.12) \qquad \sum_{i=1}^{n} X_i \exp \left\{ \int_0^{\Psi} \frac{-K_i(T(\alpha), T_i)}{1-\alpha} d\alpha \right\} = 1 - \Psi .$$

5. Continuous Distribution of Components

We can go over to the continuous case directly by assuming the distribution of components may be parameterized by a function $X_0(\alpha)$ with a corresponding boiling point function $T_0(\alpha)$. Then Equation (4.8) may be replaced by

$$(5.1) \qquad \phi = \int_0^{\alpha} X_0(\beta)\, d\beta .$$

Eliminating the parameter α gives the true boiling point curve $T = \hat{T}(\phi)$. If this function is given in the first place, then the distribution is arbitrary so we may as well choose $X_0(\alpha) = 1$. In this case Equation (4.12) becomes

$$(5.2) \qquad \int_0^1 \exp \left\{ \int_0^{\Psi} \frac{-K(T(\alpha), \hat{T}(\beta))}{1-\alpha} d\alpha \right\} d\beta = 1 - \Psi .$$

6. Equilibrium Ratio

An estimate of vapor pressures of hydrocarbons may be written in the form

$$(6.1) \qquad \log K(T, \hat{T}) = \frac{k(T-\hat{T})}{(1+ T/\tau)(1 - \hat{T}/\mu)} ,$$

where $\hat{T}$ is the normal boiling point of the particular pseudo-component (a mixture with a very narrow boiling range) and T is the temperature of the material. This represents the data summarized by the Esso Chart, page 669 in Van Winkle's text which was also published in Reference [15]. We judge the constants to be

$$(6.2)\qquad \begin{cases} k = .021,\ 1/{}^{\circ}F \\ \tau = 460,\ {}^{\circ}F \\ \mu = 3028,\ {}^{\circ}F \end{cases}$$

If it is assumed that the fugacity coefficient is one, which is probably very good for atomspheric distillations and it is assumed that the activity coefficient is one, which is probably not so good but possible for hydrocarbons, and further it is assumed that there is no non-ideal behavior due to mixtures, the ratio of vapor pressure to the system pressure may be taken as an estimate of the equilibrium ratio K. This has been already anticipated by the notation of Equation (6.1). Short of these assumptions things become exceedingly complicated as exemplified by the computations in References [16, 17].

7. Numerical Solution

Equation (5.2), with the equilibrium ratio given by Equation (6.1), is a rather peculiar functional relation between the differential distillation curve $T(\Psi)$ and the true boiling point of curve $\hat{T}(\Psi)$. We have not found this equation handled directly in the literature. Usually Ralyeigh's Equation is separated according to components. We look for a numerical procedure for estimating $T(\Psi)$ given $\hat{T}(\Psi)$. An approach which is successful is to make steps in Ψ sufficiently small so that the linear terms in the expansion for the exponential for K suffice. This allows separation of variables so that the inner integral is free of the parameter of integration of the outer integral and may be evaluated by a Newton-Raphson iteration. The boiling point curve is then recovered by a numerical differentiation.

To find the initial point of the distillation, we differentiate (5.2) to get

$$(7.1)\qquad \int_0^1 K(T(\Psi),\ \hat{T}(\beta))\ \exp\left\{\int_0^{\Psi} \frac{-K(T(\alpha),\ \hat{T}(\beta))}{1-\alpha}\, d\alpha\right\}\ d\beta = 1-\Psi .$$

Then $T(0)$ may be computed from the equation

$$(7.2) \qquad \int_0^1 K(T(0), \hat{T}(\beta))d\beta = 1.$$

The data which we use is taken from Reference [1]. The true boiling point curve is given in ten discrete steps so Simpson's rule is used for the integration on β. Newton's iteration yields $T(0)$. We attempted to used Equation (7.1) for the entire computation but the numerical scheme proved unstable.

To simplify the computations let

$$(7.3) \qquad \begin{cases} U(\alpha) = T(\alpha) + \tau \\ \hat{U}(\beta) = U(\beta) + \tau \\ g(\beta) = \dfrac{k\tau}{1 - \hat{T}(\beta)/\mu} \end{cases} .$$

Then (6.1) is rewritten as

$$(7.4) \qquad K(T(\alpha), \hat{T}(\beta)) = \exp\left\{g(\beta)\left[1 - \frac{\hat{U}(\beta)}{U(\alpha)}\right]\right\} .$$

Adding and subtracting $U(\beta)/U(\Psi_k)$ leads to

$$(7.5) \qquad K(T(\alpha), \hat{T}(\beta)) = F_k(\beta) \exp\left\{ H_k(\beta)\left[1 - \frac{U(\Psi_k)}{U(\alpha)}\right]\right\},$$

with

$$(7.6) \qquad \begin{cases} H_k(\beta) = \dfrac{g(\beta)\hat{U}(\beta)}{U(\Psi_k)} \\ F_k(\beta) = \exp\{g(\beta) - H_k(\beta)\}. \end{cases}$$

The kernel K in Equation (7.5) is approximated by the linear portion of the expansion for the exponential and substituted into the functional equation (5.2) to give

$$(7.7) \qquad \int_0^1 Q_k(\beta)\exp\left\{F_k(\beta)\left[\log\frac{1-\Psi}{1-\Psi_k} - H_k(\beta)\,\gamma_k(\Psi)\right]\right\} d\beta = 1-\Psi,$$

where

$$(7.8) \qquad \gamma_k(\Psi) = \int_{\Psi_k}^{\Psi}\left[1 - \frac{U(\Psi_k)}{U(\alpha)}\right]\frac{d\alpha}{1-\alpha},$$

$$(7.9) \qquad Q_k(\beta) = \exp\left\{\int_0^{\Psi_k}\frac{-K}{1-\alpha}\,d\alpha\right\}.$$

As in the computation for $T(0)$, $\gamma_k(\Psi)$ is found by a Newton iteration while the integral on β is evaluated by Simpson's rule. The distillation curve is found by differentiating Equation (7.8) to get

$$(7.10) \qquad T(\Psi) = -\tau + \frac{U(\Psi_k)}{1-(1-\Psi)\gamma'_k(\Psi)}.$$

Thus, the computation proceeds from $T(0)$ by choosing intervals $\Psi_1, \Psi_2, \ldots, \Psi_n = 1$ small enough for the approximation of the exponential to be valid and within these intervals by choosing steps in Ψ to allow numerical differentiation of $\gamma_k(\Psi)$ to hold. The computer program to do this is actually rather modest and our time-shared system does the computation faster than the response of the teletype output.

8. Conclusion

Examples of the calculation are shown in the figures for data from Reference 1 on samples of gasolines and oil mixtures. The differential distillation curves which we calculate do not completely match the ASTM results. Possibly this is because some reflux action occurs in the ASTM procedure.

References

1. R.L. Geddes, "Computation of Petroleum Fractionation," Ind. and Eng. Chem. Vol. 33, No. 6, June 1941, 795-801.

2. W. C. Edmister and K. I. Okamoto, "Applied Hydrocarbon Thermodynamics, Part 12: Equilibrium Flash Vaporization Calculation for Petroleum Fractions," Petroleum Refiner, Vol. 38, No. 8, August 1959, 117-129.

3. W. C. Edmister, Applied Hydrocarbon Thermodynamics, Gulf Publishing Co., Houston, 1961.

4. M. Van Winkle, "Easy Way to Figure EFV and TBP," Hydrocarbon Processing and Petroleum Refiner, Vol. 43, No. 4, April 1964, 139-142.

5. M. Van Winkle, Distillation, McGraw-Hill, 1967.

6. W. L. Nelson, Petroleum Refining Engineering, Fourth Edition, McGraw-New York, 1958, 111-120.

7. E. J. Hoffman, "Relations between true boiling point and ASTM distillation curves," Chem. Eng. Sci., 1969, Vol. 24, 113-117.

8. E. J. Hoffman, "Differential vaporization curves for complex mixtures," Chem. Eng. Sci., 1969, Vol. 24, 1736-38.

9. E. J. Hoffman, Azeotropic and Extractive Distillation, Interscience Pub. 1964, 75-89.

10. S. S. Gandbhir and P. S. Virk, "Rapid interconversion between ASTM and TBP distillations," Oil and Gas J., January 11, 1971, 53-55.

11. J. B. Maxwell, Data Book on Hydrocarbons, D. van Nostrand, 1950, pp. 222-229.

12. D. L. Taylor and W. C. Edmister, "Solutions for Distillation Processes Treating Petroleum Fractions," AIChE Journal Vol. 17, No. 6, Nov. 1971, 1324-1329.

13. O. H. Hariu and R. C. Sage, "Crude Split Figured by Computer," Hydrocarbon Proc., April 1969, 143-148.

14. J. D. Baker and E. R. Rang, "Modeling processes with integral operators: application to distillation correlation," Proceedings of the 1972 Summer Simulation Conference, June 13-16, San Diego, Calif.

15. J. B. Maxwell and L.S. Bonnell, Derivation and Precision of a New Vapor Pressure Correlation for Petroleum Hydrocarbons," Ind. Eng. Chem. Vol. 49, No. 7 (1959), 1187-1196.

16. R. A. Williams and E. J. Henley, "A Comprehensive Vapor-Liquid Equilibrium Computer Program - KVALUE," Chem. Eng. J., Vol. 1, 1970, 145-151.

17. F. O. Mixon, B. Gumowski and B. H. Carpenter, "Computation of Vapor-Liquid Equilibrium Data from Solution Vapor Pressure Measurements," Inc. and Eng. Chem. Fundamentals, Vol. 4, No. 4, Nov. 1965, 455-459.

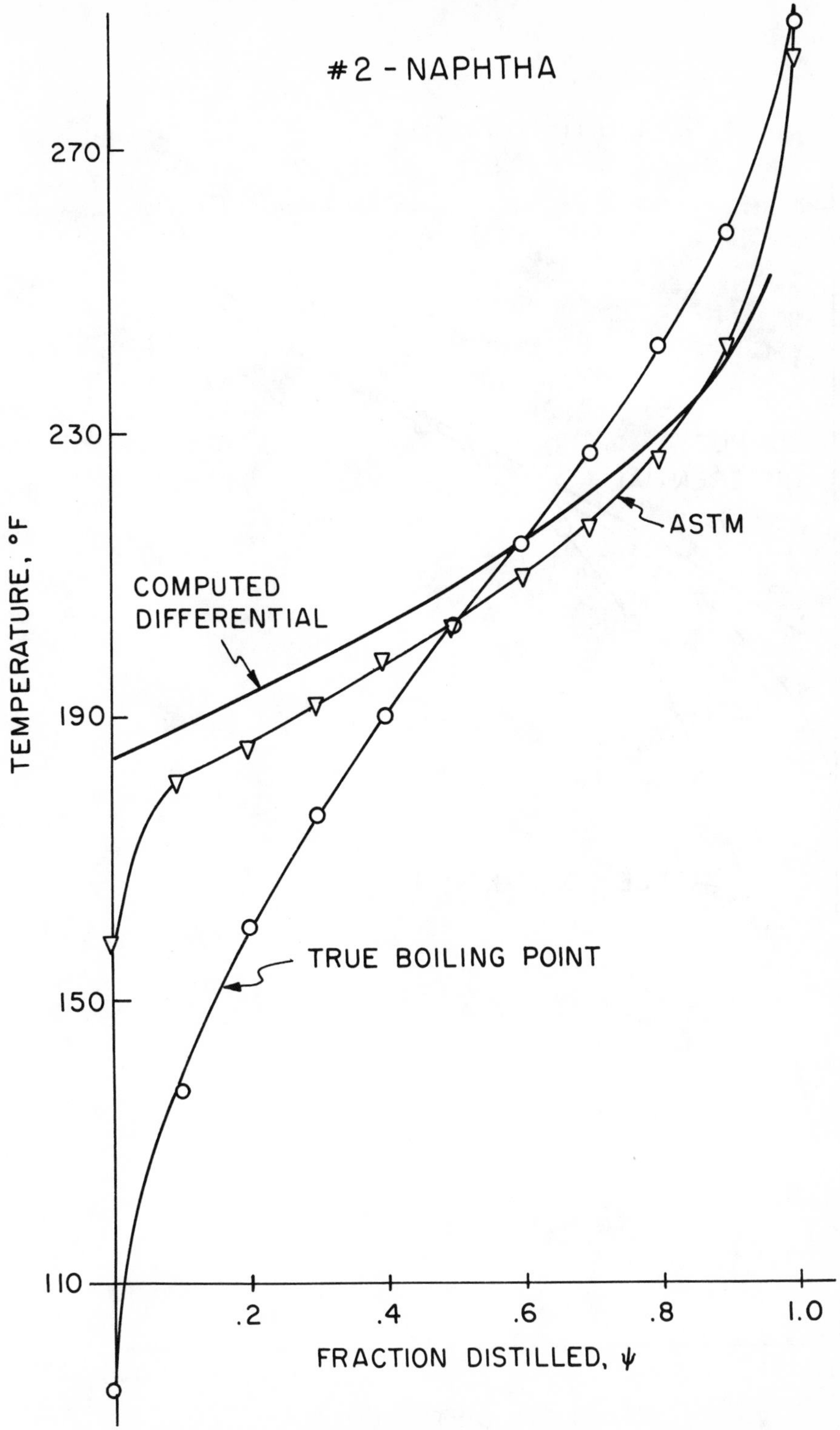

#2 - NAPHTHA
270
230
190
150
110
TEMPERATURE, °F
COMPUTED
DIFFERENTIAL
ASTM
TRUE BOILING POINT
.2
.4
.6
.8
1.0
FRACTION DISTILLED, ψ

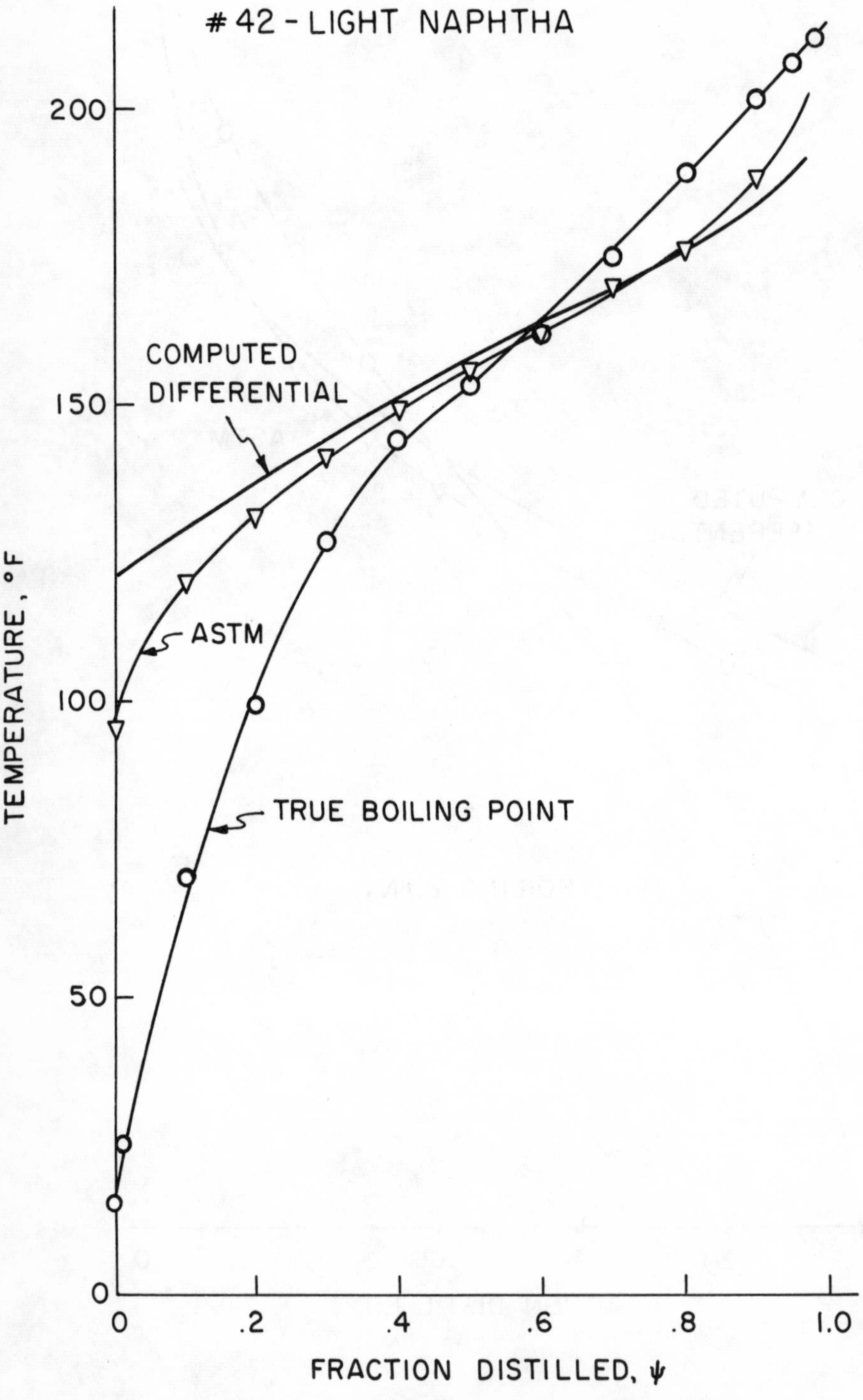

42 - LIGHT NAPHTHA
COMPUTED DIFFERENTIAL
ASTM
TRUE BOILING POINT
TEMPERATURE, °F
200
150
100
50
0
0
.2
.4
.6
.8
1.0
FRACTION DISTILLED, ψ

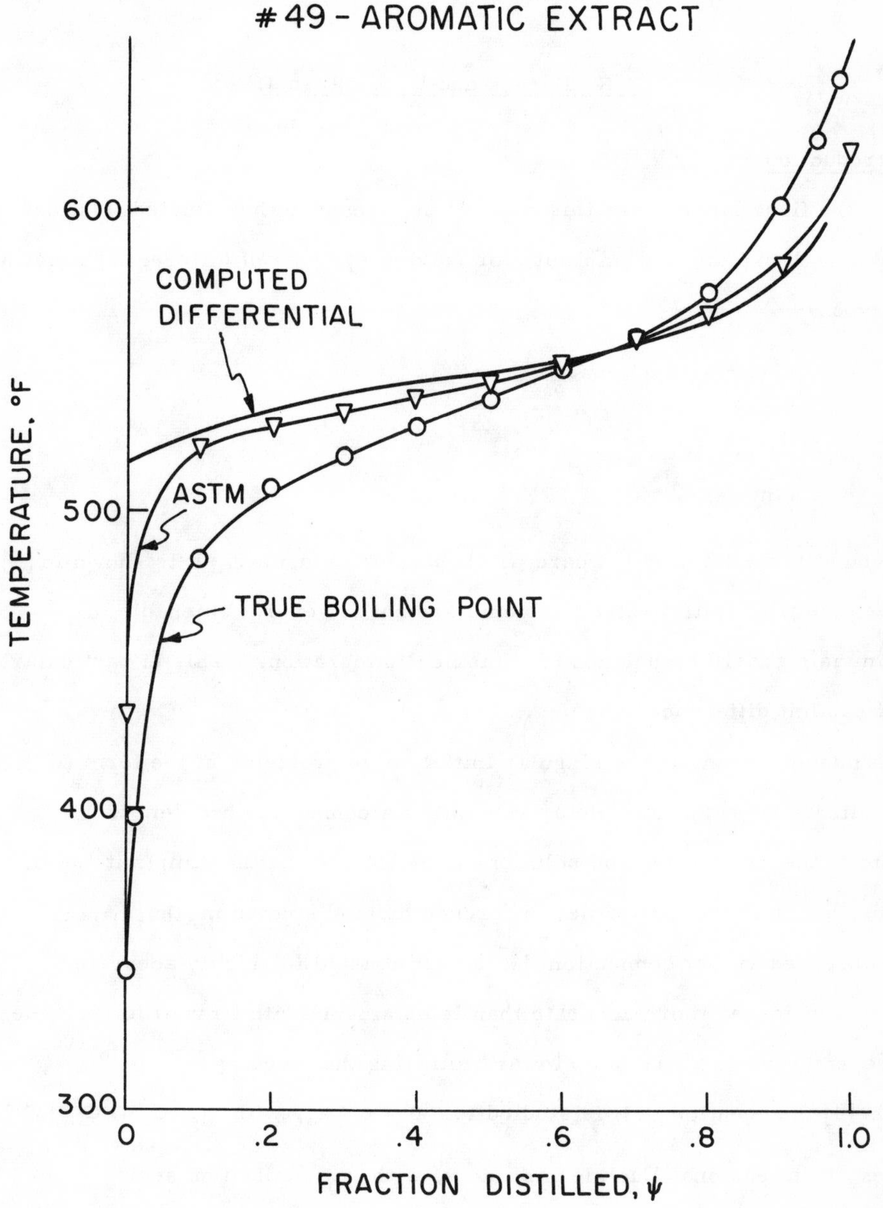

#49 - AROMATIC EXTRACT
COMPUTED DIFFERENTIAL
ASTM
TRUE BOILING POINT
TEMPERATURE, °F
600
500
400
300
0
.2
.4
.6
.8
1.0
FRACTION DISTILLED, ψ

Totally Implicity Methods for Numerical Solution of Singular Initial Value Problems

E. R. Barnes and D. L. Russell

1. Introduction

In the first paper under this title [4] the second author studied first order methods, both explicit and implicit, for solving systems of differential equations of the type

(1.1) $$x^r y' = g_0(x, y).$$

The initial conditions

(1.2) $$y(0) = 0$$

are given at the point $x = 0$ where (1.1) becomes singular. It is shown in [4] that such singular initial value problems are not in general amenable to solution via standard techniques of numerical integration. This is particularly true of explicit difference schemes.

Numerous examples of singular initial value problems of the form (1.1), (1.2) can be cited. In [4] an example arising out of a connection problem in asymptotic theory is given and solutions are obtained via the "implicit Euler's method", which is the first order difference method studied in that paper. It was observed in that connection that a need existed for highly accurate solution techniques, more accurate than is obtainable with first order schemes. The present work is a first step toward fulfilling that need.

In [4] the equation (1.1) is studied for $0 < x \leq a_0$, with g_0: $\mathbb{C}^n \to \mathbb{C}^n$ ($\mathbb{C}^n$ is complex n-dimensional Euclidean space.), with g_0 defined on some neighborhood $\|y\| \leq b_0$ of the origin in $\mathbb{C}^n$. Here a_0 and b_0 are positive and $\| \ \|$ is the usual Euclidean norm in $\mathbb{C}^n$. The results there are obtained under the assumption that the Jacobian matrix $\frac{\partial g}{\partial y}$ is continuous for $0 < x \leq a_0$, $\|y\| \leq b_0$ and that all of the eigenvalues of the matirx

$$\frac{\partial g_0}{\partial y}(0,0)$$

have negative real parts. These assumptions are used, in particular, to prove the existence of a unique solution $y(x)$ of (1.1), (1.2) in an interval $0 < x \leq a \leq a_0$.

We believe that the techniques of the present paper will eventually be shown to be applicable to initial value problems of the generality described above. But the proof of the validity of the techniques which we present involves great algebraic complexities and we must, in this paper, confine our attention to scalar equations (1.1). We shall assume g_0 to be real and of class C^∞ for $0 < x \leq a_0$, $|y| \leq b_0$ with

$$g_0(0,0) = 0, \quad \frac{\partial g_0}{\partial y}(0,0) = -\lambda, \ \lambda > 0. \tag{1.3}$$

With

$$g(x,y) = g_0(x,y) + \lambda y \tag{1.4}$$

our equation becomes

$$x^r y' = -\lambda y + g(x,y)$$

where, clearly,

$$\frac{\partial g}{\partial y}(0,0) = 0.$$

2. Totally Implicit Difference Methods

Let $h > 0$ and define

$$x_k = kh, \ k = 0,1,2,3,\ldots \ . \tag{2.1}$$

The idea behind explicit difference schemes for solving differential equations

$$y' = f(x,y)$$

is to obtain, via Lagrange interpolation, a polynomial $p(x)$ of degree $n-1$ having values $f(x, \eta_k)$ at points x_k, $k = k_0 - n+1, k_0 - n+2, \ldots, k_0$, where approximations η_k to the solution values $y(x_k)$ have already been obtained. An approximation η_{k_0+1} to $y(x_{k_0+1})$ is then obtained via the formula

$$\eta_{k_0+1} = \eta_{k_0} + \int_{x_{k_0}}^{x_{k_0+1}} p(x)\,dx. \tag{2.2}$$

Schemes which are, in the literature, called implicit, differ from this in that $p(x)$ is required to have the values $f(x_k, \eta_k)$ at the points x_k, $k = k_0 - n+2, k_0 - n+3, \ldots, k_0, k_0+1$, so that $p(x)$ itself depends upon η_{k_0+1}. The formula (2.2) is then implicit and solutions are normally obtained by iteration. It should be noted, however, that these schemes are still largely explicit in the sense that $n - 1$ of the n values of η_k used in defining $p(x)$ are already assumed known. It would therefore be reasonable to call such schemes <u>semi-implicit.</u>

By a <u>totally implicit</u> difference scheme, we mean a method whereby, given an approximation η_{k_0} to $y(x_{k_0})$, approximations $\eta_{k_0+1}, \eta_{k_0+2}, \ldots, \eta_{k_0+n}$ are obtained by requiring that $p(x)$ be a polynomial of degree $n-1$ having values $f(x_k, \eta_k)$ at x_k, $k = k_0+1, \ldots, k_0+n$, where the η_k satisfy

$$\eta_{k+1} = \eta_k + \int_{x_k}^{x_{k+1}} p(x)dx, \; k = k_0, \ldots, k_0 + n-1.$$

This involves solution of a system of n equations, nonlinear in general, in the n unknowns η_k, $k = k_0 + 1, \ldots, k_0 + n$. For "every day" solution of ordinary differential equations such a scheme would be highly impractical. But we shall see that such totally implicit schemes do make sense in the context of singular initial value problems, even though it turns out that not all of the n values η_k, $k = k_0 + 1, \ldots, k_0 + n$, calculated at each step, are actually usable.

There is a strong parallel between our methods and the "implicit Runge-Kutta" methods developed earlier by Butcher [1]. This parallel is studied in some detail by Wright [6] who establishes that implicit Runge-Kutta methods are equivalent to certain collocation schemes. The whole are of A-stable integration methods and "stiff" differential equations are very pertinent to the study of singular initial value problems. We are hoping to explore these connections in later work.

3. The Recursion Equations

Let the points x_k satisfy (2.1) and let $y(x)$ be a function known to have a bounded $(n+1)$-st derivative in an interval $0 \leq x \leq a_0$, $a_0 > 0$. We will assume the points x_k lie in this interval. The boundedness of $y^{(n+1)}(x)$ implies that

$$y'(x) = p_{k+1}(x) + h^n E_{k+1}(x)$$

where, for some fixed $B > 0$

$$(3.1) \qquad |E_{k+1}(x)| \leq B, \quad x_k \leq x \leq x_{k+n}, \quad k = 0, 1, 2, 3, \ldots,$$

and $p_{k+1}(x)$ is the polynomial of degree $n-1$ in x interpolating the function y' at the points $x_{k+1}, x_{k+2}, \ldots, x_{k+n}$. We represent $p_{k+1}(x)$ in the form

$$p_{k+1}(x) = a^1_{k+1}\left(\frac{x-x_k}{h}\right)^{n-1} + a^2_{k+1}\left(\frac{x-x_k}{h}\right)^{n-2} + \cdots + a^n_{k+1}$$

Evaluating $p_{k+1}(x)$ at $x_{k+\ell} = x_k + \ell h$, for $\ell = 1, 2, \ldots, n$, and noting that $p_{k+1}(x_{k+\ell})$ must equal $y'(x_{k+\ell})$, we obtain equations for the coefficients a^ℓ_{k+1}, $\ell = 1, 2, \ldots, n$, namely,

$$(3.2)\quad \begin{pmatrix} 1 & 1 & \dots & 1 & 1 \\ 2^{n-1} & 2^{n-2} & \dots & 2 & 1 \\ \vdots & \vdots & & \vdots & \vdots \\ n^{n-1} & n^{n-2} & \dots & n & 1 \end{pmatrix} \begin{pmatrix} a^1_{k+1} \\ a^2_{k+1} \\ \vdots \\ a^n_{k+1} \end{pmatrix} = \begin{pmatrix} y'(x_{k+1}) \\ y'(x_{k+2}) \\ \vdots \\ y'(x_{k+n}) \end{pmatrix}$$

Let $C = (c^i_j)$ denote the inverse of the Vandemorde matrix V on the left of (3.2). Then clearly

$$a^i_{k+1} = \sum_{j=1}^{n} c^i_j \, y'(x_{k+j}), \quad i = 1, 2, \dots, n,$$

so that

$$(3.3)\quad p_{k+1}(x) = \sum_{i=1}^{n} \left(\sum_{j=1}^{n} c^i_j \, y'(x_{k+j}) \right) \left(\frac{x - x_k}{h} \right)^{n-i} .$$

Since

$$(3.4)\quad y(x_{k+\ell}) - y(x_{k+\ell-1}) = \int_{x_{k+\ell-1}}^{x_{k+\ell}} y'(x)dx = \int_{x_{k+\ell-1}}^{x_{k+\ell}} p_k(x)\,dx$$

$$+ \; h^n \int_{x_{k+\ell-1}}^{x_{k+\ell}} E_{k+1}(x)dx, \quad \ell = 1, 2, \dots, n,$$

we have, after substituting (3.3) into (3.4) and performing the indicated integrations,

$$y(x_{k+\ell}) - y(x_{k+\ell-1}) = h \sum_{i=1}^{n} \left(\sum_{j=1}^{n} c^i_j y'(x_{k+j}) \right) \left(\frac{\ell^{n-i+1} - (\ell-1)^{n-i+1}}{n-i+1} \right)$$

$$+ \; h^n \int_{x_{k+\ell-1}}^{x_{k+\ell}} E_{k+1}(x)\,dx =$$

$$= h \sum_{j=1}^{n} \left(\sum_{i=1}^{n} c_j^i \left[\frac{\ell^{n-i+1} - (\ell-1)^{n-i+1}}{n-i+1} \right] \right) y'(x_{k+j})$$

(3.5) $$+ \; h^n \int_{x_{k+\ell-1}}^{x_{k+\ell}} E_{k+1}(x)dx \equiv h \sum_{j=1}^{n} R_j^\ell y'(x_{k+j}) + h^{n+1} \beta_{k+1}^\ell .$$

The definition of R_j^ℓ is clear and β_{k+1}^ℓ is a real number such that for all k and ℓ

(3.6) $$|\beta_{k+1}^\ell| \leq B \quad \text{(cf. (3.1))}.$$

Now let $y(x)$ solve (1.1), (1.2). Sufficient conditions under which $y(x)$ will have a bounded $(n+1)$ st derivative are easily given. (Some results in this direction are presented in [4].) We shall assume that these conditions are fulfilled. For $y(x)$ satisfying (1.1) we have

$$y'(x_{k+j}) = x_{k+j}^{-r} \, g_0(x_{k+j}, y(x_{k+j})), \quad j = 1, 2, \ldots, n,$$

and the equations (3.5) become, for $\ell = 1, 2, \ldots, n$,

(3.7) $$y(x_{k+\ell}) - y(x_{k+\ell-1}) = h \sum_{j=1}^{n} R_j^\ell x_{k+j}^{-r} \, g_0(x_{k+j}, y(x_{k+j})) + h^{n+1} \beta_{k+1}^\ell .$$

Let us observe now that the matrix R whose entries are

$$R_j^\ell = \sum_{i=1}^{n} c_j^i \left[\frac{\ell^{n-i+1} - (\ell-1)^{n-i+1}}{n-i+1} \right]$$

can be written in the form

$$R = LD_1^{-1} VD_2 C ,$$

where V and $C = V^{-1}$ have been described above,

$$L = \begin{pmatrix} 1 & 0 & 0 & \dots & 0 & 0 \\ -1 & 1 & 0 & \dots & 0 & 0 \\ 0 & -1 & 1 & \dots & 0 & 0 \\ \vdots & \vdots & \vdots & & \vdots & \vdots \\ 0 & 0 & 0 & \dots & 1 & 0 \\ 0 & 0 & 0 & \dots & -1 & 1 \end{pmatrix}$$

$$D_1 = \operatorname{diag}\left(1, \frac{1}{2}, \dots, \frac{1}{n}\right),$$

and

$$D_2 = \operatorname{diag}\left(\frac{1}{n}, \frac{1}{n-1}, \dots 1\right).$$

Then, since

$x^{-r}_{k+j} = \left(1 + \frac{j-1}{k+1}\right)^{-r} x^{-r}_{k+1}$, $j = 2, 3, \dots, n$, the matrix $\hat{R}_{k+1}$ whose (ℓ, j)-th entry is $R^{\ell}_{j}\, x^{-r}_{k+j}$ is given by

$$\hat{R}_{k+1} = x^{-r}_{k+1}\, L D_1^{-1} V D_2\, C D_3(k+1),$$

with

$$D_3(k+1) = \operatorname{diag}\left(1, \left(1 + \frac{1}{k+1}\right)^{-r}, \dots, \left(1 + \frac{n-1}{k+1}\right)^{-r}\right). \tag{3.8}$$

Thus (3.7) can be written in the form

$$L\, y_{k+1} = h x^{-r}_{k+1}\, \hat{R}_{k+1}\, G_0(x_{k+1}, y_{k+1}) + y^{1}_{k}\, e_1 + h^{n+1} \boldsymbol{\beta}_{k+1} \tag{3.9}$$

where

$$(3.10)\quad y_{k+1} = \begin{pmatrix} y(x_{k+1}) \\ y(x_{k+2}) \\ \vdots \\ y(x_{k+n}) \end{pmatrix}, G_0(x_{k+1}, y_{k+1}) = \begin{pmatrix} g_0(x_{k+1}, y(x_{k+1})) \\ g_0(x_{k+2}, y(x_{k+2})) \\ \vdots \\ g_0(x_{k+n}, y(x_{k+n})) \end{pmatrix}, \quad e_1 = \begin{pmatrix} 1 \\ 0 \\ \vdots \\ 0 \end{pmatrix}.$$

Our numerical method involves the sucessive calculation of vectors

$$\eta_k = \begin{pmatrix} \eta_k^1 \\ \eta_k^2 \\ \vdots \\ \eta_k^n \end{pmatrix}, \; k = 1, 2, 3, \dots,$$

whose first components η_k^1 will be approximation to $y_k^1 = y(x_k)$. With $G_0(x_{k+1}, \eta_{k+1})$ defined just as in (3.10) but with $g_0(x_{k+j}, y(x_{k+j}))$ replaced by $g_0(x_{k+j}, \eta_{k+j}^j)$, the vectors η_k are required to satisfy

$$(3.11)\qquad L\eta_{k+1} = hx_{k+1}^{-r}\ \hat{R}_{k+1}\ G_0(x_{k+1}, \eta_{k+1}) + \eta_k^1\ e_1 .$$

(Note that only one component of η_k enters into the computation of η_{k+1}.)

The equations (3.9), (3.11) can also be written as

$$D_1\, y_{k+1} - hx_{k+1}^{-r}\ VD_2CD_3(k+1)\ G_0(x_{k+1}, y_{k+1})$$

$$(3.12)\qquad = y_k^1\ e_1 + h^{n+1}\ \hat{\boldsymbol{\beta}}_{k+1},$$

where

$$(3.13)\qquad \hat{\boldsymbol{\beta}}_{k+1} = D_1\, L^{-1}\, \boldsymbol{\beta}_{k+1}$$

and

$$D_1 \eta_{k+1} - hx_{k+1}^{-r} VD_2 CD_3(k+1) G_0(x_{k+1}, \eta_{k+1})$$

(3.14) $$= \eta_k^1 e_1 ,$$

respectively. Here we have used the fact that

$$D_1 L^{-1} e_1 = e_1 .$$

4. A Problem in Linear Algebra

From (1.4) and (3.10) we see that, for x_{k+1} and y_{k+1} small, $G_0(x_{k+1}, y_{k+1})$ is approximately equal to $- \lambda y_{k+1}$. Also, (3.8) shows that

(4.1) $$\lim_{k \to \infty} D_3(k+1) = I.$$

The form of the recusion equations(3.12), (3.14) therefore indicates that the matrix

$$D_1 + \lambda hx_{k+1}^{-r} VD_2 C$$

will be of some importance. When $r \neq 1$ the expression λhx_{k+1}^{-r} may take on any values between 0 and $+ \infty$, depending upon the relationship between k and h. Consequently we wish to study the matrix function

$$Q(s) = D_1 + s VD_2 C$$

on the interval $0 < s < 1$ and we are particularly concerned with the existence and uniform boundedness of $Q(s)^{-1}$ there.

Let

$$q(s) = \det Q(s)$$

be the n-th degree characteristic polynomial of $Q(s)$ in each dimension $n=1, 2, 3, \ldots$. Letting

(4.2) $$P(s) = Q(s)^{-1}$$

wherever the latter exists, we apply Cramer's rule to see that the first diagonal entry $p_1^1(s)$ of $P(s)$ has the form

$$(4.3) \qquad p_1^1(s) = \frac{\tilde{q}(s)}{q(s)} ,$$

where $q(s)$ is as described above and $\tilde{q}(x)$ is a polynomial

$$\tilde{q}(s) = \sum_{i=0}^{n-1} q_i s^i$$

of degree n-1 in s. Similarly we have

$$q(s) = \sum_{i=0}^{n} q_i s^i .$$

__Lemma A.__ __For any positive integer__ $n = 1, 2, 3, \ldots$ __the coefficients__ q_i __of__ $q(s)$ __are all positive__. __Moreover, the coefficients__ $\tilde{q}_i$ __of__ $\tilde{q}(s)$ __satisfy__

$$\tilde{q}_0 = q_0 ,$$

$$0 < \tilde{q}_i < q_i , \quad i = 1, 2, \ldots, n-1, \quad q_n > 0 .$$

__Proof.__ Following a method developed in [2], [3], [6], one can express $p_i^i(s)$ in the following form for $Re(s) < 0$:

$$p_i^i(s) = \frac{\int_0^\infty e^{s\sigma} \prod_{j=1}^{n} (\sigma + 1 - j)\, d\sigma}{\int_0^\infty e^{e\sigma} \prod_{j=1}^{n} (\sigma - j)\, d\sigma} .$$

Let us write

$$\prod_{j=1}^{n} (\sigma + 1 - j) = \sum_{j=0}^{n} (-1)^{n-j} v_j \sigma^j$$

$$\prod_{j=1}^{n} (\sigma - j) = \sum_{j=0}^{n} (-1)^{n-j} u_j \sigma^j .$$

It is clear then that $v_n = u_n = 1$, $v_0 = 0$, and the remaining v_j are intergers with

(4.4) $\quad 0 < v_j < u_j$, $j = 1, 2, \ldots, n-1$

But, for $\mathrm{Re}(s) < 0$, using integration by parts,

$$\int_0^{\infty} e^{s\sigma} \sigma^i \, d\sigma = (-1)^j {}_j! \; s^{-(j+1)}.$$

Thus for $\mathrm{Re}(s) < 0$ we obtain

$$p_i^i(s) = \frac{\sum_{j=0}^{n} (-1)^{n-j} v_j (-1)^j {}_j! \; s^{-(j+1)}}{\sum_{j=0}^{n} (-1)^{n-j} u_j (-1)^j {}_j! \; s^{-(j+1)}}$$

$$= \frac{\sum_{i=0}^{n} (n-i)! \; v_{n-i} \, s^i}{\sum_{i=0}^{n} (n-i)! \; u_{n-i} \, s^i} .$$

Since $p_i^i(s)$ is a rational function, this formula for $p_i^i(s)$ must, in fact, hold true for all s. It follows that there is a positive rational number $\hat{q}$ such that

$$q_i = \hat{q}(n-i)! \; u_{n-i}, \; i = 0, 1, 2, \ldots, n$$

$$\tilde{q}_i = \hat{q}(n-i)! \; v_{n-i}, \; i = 0, 1, 2, \ldots, n-1.$$

Together with (4.4) and the remarks immediately preceding it, this completes the proof of Lemma A.

Remark As a consequence of this lemma, we see that the polynomial $q(s)$ has no zeros on the half line $s \geq 0$ and $q(s) \geq \tilde{q}(s)$ there. In fact, it is easy to see that there are positive numbers K, $\tilde{K}$ such that

(4.5) $\|P(s)\| \leq \frac{K}{1+s}$, $0 < s < \infty$

(4.6) $0 < p_1^1(s) < \frac{1}{1+\tilde{K}s}$, $0 < s < \infty$

In a later paper we intend to show that our methods also apply to yield approximate solutions for the initial value problem (1.1), (1.2) when y is an n-vector and the Jacobian matrix $\frac{\partial y}{\partial y}(0,0)$ has only eigenvalues with negative real parts. In order to carry this program out we will have to establish the following result, which we now state as a conjecture. It can be verified quite readily for $n = 1, 2$.

Conjecture. *The polynomials $q(s)$ have only zeros with negative real parts for any value of n and the rational functions $p_1^1(s)$ map the half plane $Re(s) \geq 0$ into a subset of the disc $|p| \leq 1$.*

5. Computational Solution of the Recursion Equations

In this section we study the question of generating vector solutions η_{k+1} of the recursion equation (3.14) for x_{k+1} and η_k^1 sufficiently small. We postpone to Section 6 the question of the validity of the numbers η_k^1 as approximations to $y_k^1 = y(x_k)$.

First of all, we set (cf. (1.4), (3.10))

$$G_0(x_{k+1}, \eta_{k+1}) = -\lambda\eta_{k+1} + G(x_{k+1}, \eta_{k+1}).$$

Then we can prove :

Theorem 1. *There exist positive constants $a, \hat{b}, \hat{h}$ and b such that if $r \neq 1$ and*

$$0 < x_{k+1} \leq a, \quad |\eta_k^1| \leq \hat{b}, \quad 0 < h \leq \hat{h},$$

then the system of equations (3.14) has a unique solution η_{k+1} in the neighborhood

$$\|\eta\| \leq b$$

of the origin in R^n and

$$(5.1) \qquad \eta_{k+1} = \lim_{\ell \to \infty} \zeta_\ell$$

where the vectors ζ_ℓ satisfy

$$\zeta_0 = 0,$$

and one of the (equivalent) recursion relations

$$(5.2) \qquad [D_1 + \lambda hx_{k+1}^{-r} VD_2 CD_3(k+1)] \zeta_{\ell+1}$$

$$= hx_{k+1}^{-r} VD_2 CD_3(k+1) G(x_{k+1}, \zeta_\ell) + \eta_k^1 e_1$$

or

$$(5.3) \qquad [D_1 + \lambda hx_{k+1}^{-r} VD_2 C] \zeta_{\ell+1} =$$

$$hx_{k+1}^{-r} VD_2 C[\lambda(I - D_3(k+1)) \zeta_\ell + G(x_{k+1}, \zeta_\ell)] + \eta_k^1 e_1 .$$

The equations (5.3) are used for $k \geq k_0$, k_0 to be defined below, and the equations (5.2) are used for $k < k_0$. The convergence (5.1) for (5.3) is independent of the conditions $r \neq 1$, $h \leq \hat{h}$, but these conditions are used to establish (5.1) for the equations (5.2).

Proof From (4.2), (4.5) we have

$$(5.4) \quad \|(D_1 + \lambda hx_{k+1}^{-r} VD_2 C)^{-1}\| \equiv \|P(\lambda hx_{k+1}^{-r})\| \leq \frac{K}{1 + \lambda hx_{k+1}^{-r}} < K.$$

We rewrite (5.3) as

(5.5) $\zeta_{\ell+1} = P(\lambda h x_{k+1}^{-r}) \left\{ h x_{k+1}^{-r} VD_2 C[\lambda(I-D_3(k+1))\zeta_\ell + G(x_{k+1}, \zeta_\ell)] + \eta_k^1 e_1 \right\}$

and suppose that

(5.6) $$|\eta_k^1| \leq \hat{b},$$

(5.7) $$\| I - D_3(k+1) \| \leq \delta_1(k+1)$$

(5.8) $$\begin{aligned} \| G(x_{k+1}, \zeta_\ell) \| &= \| G(x_{k+1}, 0) + G(x_{k+1}, \zeta_\ell) - G(x_{k+1}, 0) \| \\ &\leq \delta_2(x_{k+1}) + \delta_3(x_{k+1}, \zeta_\ell) \| \zeta_\ell \|, \end{aligned}$$

(5.9) $$\| VD_2 C \| = K_0.$$

Then (5.4) - (5.9) yield

(5.10) $$\begin{aligned} \| \zeta_{\ell+1} \| &\leq \frac{K}{1 + \lambda h x_{k+1}^{-r}} K_0 \, h x_{k+1}^{-r} \, [\lambda \delta_1(k+1) \| \zeta_\ell \| \\ &\quad + \delta_2(x_{k+1}) + \delta_3(x_{k+1}, \zeta_\ell) \| \zeta_\ell \|] + K \hat{b} \\ &\leq KK_0 \left(\delta_1(k+1) + \frac{\delta_3(x_{k+1}, \zeta_\ell)}{\lambda} \right) \| \zeta_\ell \| \\ &\quad + KK_0 \frac{\delta_2(x_{k+1})}{\lambda} + K\hat{b} \end{aligned}$$

From our assumptions on the function $g(x, y)$ (See Section 1) and (4.1), it is clear that by choosing a and b sufficiently small, k_0 sufficiently large, and requiring

(5.11) $$0 < x_{k+1} \leq a, \quad \| \zeta_\ell \| \leq b, \quad k \geq k_0$$

we can ensure that

(5.12) $$KK_0 \left(\delta_1(k+1) + \frac{\delta_3(x_{k+1}, \zeta_\ell)}{\lambda} \right) \leq \gamma < 1.$$

Then (5.10) gives

$$\|\zeta_{\ell+1}\| \leq \gamma \; \|\zeta_\ell\| + KK_0 \; \frac{\delta_2(x_{k+1})}{\lambda} + K\hat{b} \, .$$

By further restricting a, if necessary, and taking $\hat{b}$ sufficiently small, we can make certain that

$$KK_0 \; \frac{\delta_2(x_{k+1})}{\lambda} + K\hat{b} < (1-\gamma)\, b.$$

Then for x_{k+1}, ζ_ℓ and k satisfying (5.11) we have

$$\|\zeta_{\ell+1}\| \leq \gamma b + (1-\gamma)\, b = b.$$

Since the condition (5.12) is easily seen to show that the mapping represented by the right hand side of (5. 5) is a contraction in the region $\|\zeta\| \leq b$, the contraction fixed point theorem shows that if successive values of ζ_ℓ are generated via (5. 3) then

$$\lim_{\ell \to \infty} \zeta_\ell \equiv \eta_{k+1}$$

exists and is the unique solution of (3.14) in the region $\|\eta\| \leq b$.

For values of $k < k_0$ we cannot state that $\|I - D_3(k+1)\| = \delta_1(k+1)$ is small and the problem of finding solutions for (3.14) must be approached differently, namely via (5. 2). We observe that (3.8) shows that

$$\|D_3(k+1)^{-1}\| = \left(1 + \frac{n-1}{k+1}\right)^r \leq n^r, \quad k \geq 0$$

and

$$\lambda h x_{k+1}^{-r} = \lambda h (k+1)^{-r} h^{-r}$$

tends uniformly $0(+\infty)$ as $h \to 0$, provided $k < k_0$ and $r < 1$ $(r > 1)$. For $0 \leq k < k_0$, therefore, we can ensure that

$$(5.13)\quad \|[D_1+\lambda h x_{k+1}^{-r}\, VD_2CD_3(k+1)]^{-1} h x_{k+1}^{-r}\, VD_2CD_3(k+1)\| \le \hat{K}\,,$$

for some fixed positive number $\hat{K}$, by requiring $0 < h < \hat{h}$, for $\hat{h}$ sufficiently small. Indeed, we can make $\hat{K}$ as close as we wish to $0(\frac{1}{\lambda})$ by taking by taking $\hat{h}$ small, again for $r < 1$ $(r > 1)$.

The recursion equation (5.2) may be rewritten

$$\zeta_{\ell+1} = [D_1+\lambda h x_{k+1}^{-r}\, VD_2CD_3(k+1)]^{-1}\left\{ h x_{k+1}^{-r}\, VD_2CD_3(k+1)G(x_{k+1}, \zeta_\ell) + \eta_k^1\, e_1\right\}$$

Combining (5.13) with a further estimate

$$\|[D_1 + \lambda h x_{k+1}^{-r}\, VD_2CD_3(k+1)]^{-1}\| \le K_1\,,$$

which one easily obtains for some positive K_1 for $0 \le k < k_0$, $0 < h \le \hat{h}$, one has (cf. (5.8))

$$\|\zeta_{\ell+1}\| \le \hat{K}\,(\delta_2(x_{k+1}) + \delta_3(x_{k+1}, \zeta_\ell))\ \|\zeta_\ell\| + K_1\,\hat{b}\,.$$

By further restructing a, b, $\hat{b}$, if necessary one can ensure

$$\hat{K}(\delta_2(x_{k+1}) + \delta_3(x_{k+1}, \zeta_\ell)) \le \gamma < 1,\quad K_1\hat{b} \le \gamma b,$$

for $0 < x_{k+1} \le a$, $\|\zeta_\ell\| \le b$. The proof that $\lim\limits_{\ell \to \infty} \zeta_\ell = \eta_{k+1}$ then proceeds as before.

Remarks. Essentially the same proof as we used for establishing the convergence of (5.3) for $k \ge k_0$ can also be used to show that if k_0 is chosen sufficiently large, then (5.2) will also be convergent. Thus (5.2) can, in fact, be used for all k, provided $0 < h < \hat{h}$, $r \neq 1$. As k_0 may be difficult to find in practice, this may well be the method to use in many cases. However, (5.3) has some advantages, provided k_0 can be estimated. The matrix P(s) can be written out explicitly as a matrix whose entries are rational functions of s similar to (4.3). The matrix $P(\lambda h x_{k+1}^{-r})$ can then be found simply by setting $s = \lambda h x_{k+1}^{-r}$ in these expressions rather than by inverting $Q(\lambda h x_{k+1}^{-r})$. This might will result in a significant reduction of computation time if h is very small and η_{k+1} is thus to be calculated for very many values of k.

When $n = 1$, $D_3(k+1) \equiv 1$ and the equation (5.2) coincides with (5.3). This is the "implicit Euler method" discussed in [4]. In this case the assumption $r \neq 1$ is not required at any point. It is probable, in fact, that this assumption is not really required at all. To do away with it, however, would require that we be able to obtain a uniform bound on

$$\|[D_1 + sVD_2CD_3(k+1)]^{-1}\|,$$

valid for $0 < s < \infty$ and $k = 0, 1, 2, 3, \ldots$.

In any event, a restriction to $r \neq 1$ is not unduly crippling. For if $r = 1$ and $g_0(x, y)$ is an analytic function of x and y, the solution $y(x)$ of (1.1), (1.2) will have a convergent power series development

$$y(x) = \sum_{j=1}^{\infty} c_j x^j$$

whose coefficients c_j may be found via recursion formulae using the Taylor series development of $g_0(x, y)$. One may then set

$$\eta_k^1 = \sum_{j=1}^{n} c_j (kh)^j, \quad 0 \leq k < k_0, \; h > 0,$$

incurring an error

$$|\eta_k^1 - y(x_k)| \leq \beta (kh)^{n+1} \leq \beta k_0^{n+1} h^{n+1}$$

for those values of k, where β is an approxpriate positive number. Having thus obtained approximations for $y(x_k)$ for $k < k_0$, the numerical procedure represented by (5.3) is used to obtain η_k^1, $k \geq k_0$.

Even when $r > 1$ it may be desirable to use a variant of this idea to replace the use of (5.2) for $k < k_0$. For $r > 1$ power series solutions (5.13) of (1.1), (1.2) do not in general exist but it is possible (see e.g. [5]) to obtain asymptotic expressions. For each integer $n > 0$ one can find a function $y_n(x)$, expressed in terms of elementary functions, such that

$$| y(x) - y_n(x) | \leq B_n x^n$$

in some interval $0 \leq x \leq a_n$. One may then define

$$\eta_k^1 = y_n(x_k), \quad k < k_0$$

and use (5.3) to generate the η_k for $k \geq k_0$.

6. Convergence of the Method

Since the vectors y_k and η_k satisfy (3.12) and (3.14), respectively, we can set

$$w_k = \eta_k - y_k, \quad k = 0, 1, 2, 3, \ldots$$

and obtain

$$(6.1) \qquad [D_1 + \lambda h x_{k+1}^{-r} \; V D_2 C D_3 (k+1)(-\tfrac{1}{\lambda} \Delta \; G_0(x_{k+1}, y_{k+1}, w_{k+1}))] \; w_{k+1}$$

$$\equiv Q(x_{k+1}, y_{k+1}, w_{k+1}) \; w_{k+1} = w_k^1 \, e_1 - h^{n+1} \hat{\beta}_k$$

where $\Delta G(x_{k+1}, y_{k+1}, w_{k+1})$ is the $n \times n$ diagonal matrix whose j-th diagonal entry is

$$\int_0^1 \frac{\partial g_0}{\partial y} (x_{k+j}, y_{k+1}^j + \tau w_{k+1}^j) \, d\tau .$$

Since

$$\frac{\partial g_0}{\partial y} (0, 0) = - \lambda ,$$

given any $\varepsilon > 0$ there are positive numbers $a(\varepsilon)$, $b(\varepsilon)$ and a positive integer $k(\varepsilon)$ such that

$$(6.2) \qquad \| D_3 (k+1) \, (- \frac{1}{\lambda} \, \Delta \, G_0(x_{k+1}, y_{k+1}, w_{k+1})) - I \; \| \leq \varepsilon$$

when

$$(6.3) \qquad 0 < x_{k+1} \leq a(\varepsilon) , \quad \| y_{k+1} \| \leq b(\varepsilon), \quad \| w_{k+1} \| \leq b(\varepsilon), \quad k \geq k(\varepsilon).$$

We again set $s = \lambda h x_{k+1}^{-r}$ and let

$$p_1^1(x_{k+1}, y_{k+1}, w_{k+1}, s)$$

(6.4)

$$= \frac{\tilde{q}(x_{k+1}, y_{k+1}, w_{k+1}, s)}{q(x_{k+1}, y_{k+1}, w_{k+1}, s)}$$

be the rational function of s which is the element in the first row and first column of the matrix Q appearing on the left hand side of (6.1). The numerator is a polynomial of degree $n-1$ in s whose first coefficient agrees with the first coefficient of $\tilde{q}(s)$ in (4.3) and the denominator is a polynomial of degree n in s whose first coefficient agrees with the first coefficient of $q(s)$ in (4.3). Assuming (6.2), (6.3) satisfied, as we make ε progressively smaller the remaining coefficients of $\tilde{q}(x_{k+1}, y_{k+1}, w_{k+1}, s)$ and $q(x_{k+1}, y_{k+1}, w_{k+1}, s)$ will tend to the corresponding coefficients of $\tilde{q}(s)$ and $q(s)$, respectively. Using Lemma A we see that for ε sufficiently small

$$| p_1^1(x_{k+1}, y_{k+1}, w_{k+1}, s) | \leq 1, \quad 0 \leq s < \infty .$$

But (6.1) clearly implies that

$$w_{k+1}^1 = p_1^1(x_{k+1}, y_{k+1}, w_{k+1}, s)\, w_k^1 - h^{n+1}\, p(x_{k+1}, y_{k+1}, w_{k+1})^T\, \hat{\beta}_{k+1}$$

where p^T is the first row of the inverse of the matrix Q defined in (6.1). If follows that (cf. (3.6), (3.13))

$$|w_{k+1}^1| \leq |w_k^1| + h^{n+1} \|p^T\| B.$$

Since the y_k are related to the actual solution $y(x)$ via (3.10), $\|y_{k+1}\|$ can be made as small as we wish by taking x_{k+1} small. Also, an examination of the proof of Theorem 1 shows that by taking k sufficiently large, x_{k+1} sufficiently small and $\hat{b}$ sufficiently small we can take the number b which appears there as small as we please.

Since

$$\|w_{k+1}\| = \|\eta_{k+1} - y_{k+1}\| \leq b + \|y_{k+1}\|,$$

we can make $\|w_{k+1}\| \leq b(\varepsilon)$ by taking b and x_{k+1} small. These observations combine to provide a proof for the following lemma.

Lemma B. If the positive numbers a, $\hat{b}$ of Theorem 1 are chosen sufficiently small, and if k_0 is chosen sufficiently large, then

$$(6.5) \qquad |\eta^1_{k+1} - y^1_{k+1}| \leq |\eta^1_k - y^1_k| + h^{n+1} P_0 B, \quad k \geq k_0,$$

where B is given by (3.6) and

$$P_0 = \sup_{\substack{k \geq k_0, \\ 0 < x \leq a \\ \|y_{k+1}\| \leq b(\varepsilon) \\ \|w_{k+1}\| \leq b(\varepsilon)}} \|p(x_{k+1}, y_{k+1}, w_{k+1})\|.$$

With the aid of Lemma B we can prove the following theorem which establishes the validity of the numbers η^1_k as approximations to $y^1_k = y(x_k)$.

Theorem 2. There exist positive numbers a, $\hat{h}$ and $\hat{B}$ such that if

$$(6.6) \qquad 0 < x_k \leq a,$$

$$(6.7) \qquad 0 < h \leq \hat{h}$$

then the vectors η_k may all be found by the methods of Theorem 1 and the numbers η^1_k, satisfy

$$(6.8) \qquad |\eta^1_k - y(x_k)| \leq \hat{B}\, x_k\, h^n.$$

Proof. Let us assume for the moment that the vectors η_k can all be found by the methods of Theorem 1, that the numbers η_k^1 all satisfy $|\eta_k^1| \leq \hat{b}$, that $\|\eta_k\| \leq b$, and that a and b (hence $\hat{b}$) have been taken small enough so that Lemma B and the points made in the discussion which led to the proof of that lemma are all valid. The inequality (6.5) will then apply for $k \geq k_0$. What we need, then, is some control over the growth of $|\eta_{k+1}^1 - y_{k+1}^1|$ for $k < k_0$. First consider the case $r > 1$. In this case we have

$$\lim_{h \to 0} hx_{k+1}^{-r} = \infty$$

uniformly for $k < k_0$. Then since the matrix $-\frac{1}{\lambda} \Delta G_0(x_{k+1}, y_{k+1}, w_{k+1})$ can be assumed as close as we wish to the identity matrix by taking $x_{k+1}, y_{k+1}, w_{k+1}$ small, we have

$$[D_1 + \lambda hx_{k+1}^{-r} VD_2CD_3(k+1)(-\tfrac{1}{\lambda}\Delta G_0(x_{k+1}, y_{k+1}, w_{k+1}))]^{-1}$$

$$(6.9) \qquad = \frac{x_{k+1}^r}{\lambda h} [(VD_2CD_3(k+1))^{-1} + E(x_{k+1}, y_{k+1}, w_{k+1}) + O(\frac{x_{k+1}^r}{\lambda h})],$$

where $E(x_{k+1}, y_{k+1}, w_{k+1})$ can be made small by choosing $x_{k+1}, y_{k+1}, w_{k+1}$ small and the term $O(\frac{x_{k+1}^r}{\lambda h})$ can be made small by taking h small, both applying uniformly for $k < k_0$. It follows therefore that by taking $x_{k+1}, y_{k+1}, w_{k+1}$, and h all sufficiently small we can guarantee that the norm of the matrix occurring on the left hand side of (6.9) is ≤ 1. Then it is an easy matter to see that (6.5) also applies for $k < k_0$.

If $r < 1$ we have

$$\lim_{h \to 0} hx_{k+1}^{-r} = 0$$

uniformly for $k < k_0$. It is then quite easy to show that the ratio (6.4) must be < 1 for h sufficiently small and for $k < k_0$ just by observing that the first diagonal entry of D_1 is unity and the first diagonal entry of the matrix

$\lambda V D_2 \, C D_3(k+1) \, (-\frac{1}{\lambda} \, \Delta G_0(x_{k+1}, y_{k+1}, w_{k+1}))$ can be confined to a compact subinterval of $(0, \infty)$ for $x_{k+1}, y_{k+1}, w_{k+1}$ sufficiently small, $k < k_0$. Thus again (6.5) holds for $k < k_0$ as well as $k \geq k_0$.

To summarize, when $r \neq 1$, $r > 0$, h is sufficiently small and $x_{k+1} \in [0, a]$, a sufficiently small,

$$(6.10) \qquad |\eta^1_{k+1} - y^1_{k+1}| \leq |\eta^1_k - y^1_k| + h^{n+1} P_0 B$$

for $k \geq 0$ so that we have, for $x_k \in [0, a]$,

$$|\eta^1_k - y^1_k| \leq k h^{n+1} P_0 B = P_0 B \, x_k h^n .$$

When $r = 1$ we cannot proceed as above. In this case, however, the method outlined at the end of Section 5 allows us to assume that for $k < k_0$

$$|\eta^1_k - y^1_k| \leq B_{n+1} x_k^{n+1} < B_{n+1} (k_0)^{n+1} h^{n+1} .$$

We then employ our numerical technique for $k \geq k_0$, for which (6.5) still applies even when $r = 1$ and obtain

$$|\eta^1_k - y^1_k| \leq B_{n+1}(k_0)^{n+1} h^{n+1} + (k - k_0) \, h^{n+1} P_0 B$$

$$(6.11) \qquad \leq \max \, (P_0 B, \, B_{n+1}(k_0)^n) \, x_k h^n .$$

Thus, taking $\hat{B} = P_0 B$, or $\max \, (P_0 B, \, B_{n+1}(k_0)^n)$, we have inequality (6.8) of Theorem 2, provided, as we have been assuming that we can keep the vectors η_k small.

Now, let us note that if we require (6.6) and (6.7) to hold and take a and h sufficiently small, we can ensure that

$$(6.12) \qquad |y^1_k| + \hat{B} x_k h^{n+1} \leq \hat{b} ,$$

The proof can then be completed rather easily by induction. We assume that (6.8) holds for a particular value of k. With a and $\hat{h}$ suitably restricted, (6.12) implies $|\eta_k^1| \leq \hat{b}$. With appropriate choices of $\hat{b}$, a and $\hat{h}$, Theorem 1 applies to give $\|\eta_{k+1}\| \leq b$. The proof of Theorem 1 requires that b be sufficiently small, requiring in turn that $\hat{b}$, and hence a and $\hat{h}$, be sufficiently small. In addition, these quantities may have to be further restricted to guarantee that Lemma B and the inequalities (6.2), (6.3) all hold assuming this to be true, the portion of the proof of Theorem 2 completed above shows that (6.8) holds also for $k+1$, and the above reasoning is repeated to get $|\eta^1_{\ell+1}| \leq \hat{b}$, $\|\eta_{\ell+2}\| \leq b$, etc. This can be continued as long as we are dealing with points $0 < x_1 \leq \cdots x_\ell \cdots \leq x_{k+1} \leq a$ and the proof of Theorem 2 is thus complete.

We have, therefore, shown the following in this paper:

(a) For $\lambda > 0$, the solution $y(x)$ of initial value problem (1.1), (1.2) can be approximated at points $x_k = kh$ in the interval $[0, a]$, provided h and a are sufficiently small, by numbers η_k^1. These numbers may be generated via (5.3) for $k \geq k_0$, k_0 remaining fixed as $h \to 0$. For $k < k_0$, the recursion equation (5.2) may be used if $r \neq 1$ but special methods may be required if $r = 1$.

(b) This numerical method is accurate to n-th order, as expressed by inequality (6.8) of Theorem 2.

References

1. J. C. Butcher: "Implicit Runge-Kulta processes", Math. Comp. 18(1964), 50-64.

2. C. Lanczos: "Trigonometric interpolation of empirical and analytical functions," J. Math. Phys. 17 (1938), 123-199.

3. C. Lanczos: "Tables of Chebyshev Polynomials," (Introduction), Nat. Bur. Stand. Appl. Math. Ser. 9 (1952).

4. D. L. Russell: "Numerical solution of singular initial value problems," SIAM J. Num. Anal. 7 (1970), 399-417.

5. W. R. Wason: Asymptotic Expansions for Ordinary Differential Equations, Interscience Pub., New York 1965.

6. K. Wright: "Some relationships between implicit Runge-Kutta, collocation and Lanczos τ methods, and their stability properties" BIT 10 (1970), 217-227.

Dichotomies for Differential and Integral Equations

George R. Sell

1. Introduction

The theory of exponential dichotomies for linear differential equations plays an important role in the study of the qualitative properties of such equations. Consider the following situation, for example.

The differential equation $x' = ax$, where $x \in R^n$ and a is an $n \times n$ matrix, is said to admit an exponential dichotomy on R^n if there exist projections P_1 and P_2 on R^n and positive constants K_1, K_2, m_1, m_2 such that

$$P_1 + P_2 = I, \quad P_1P_2 = P_2P_1 = 0$$

$$|e^{at}P_1| \leq K_1 e^{m_1 t}, \quad t \leq 0$$

$$|e^{at}P_2| \leq K_2 e^{-m_2 t}, \quad t \geq 0 .$$

We know [1] that if the eigenvalues of the matrix a have nonzero real parts, then the differential equation $x' = ax$ admits an exponential dichotomy. Moreover it is known that if $x' = ax$ admits an exponential dichotomy and if $f(t)$ is bounded on R, then the inhomogeneous equation

$$x' = ax + f \tag{1}$$

has a solution y that is bounded on R. Furthermore this solution can be represented in the form

---

This research was done in part while the author was visiting Il Istituto Matematico, Universita degli Studi di Firenze under auspices of the Italian Research Council(C.N.R). Partial support was also obtained from NSF Grant No. 27275.

$$y(t) = \int_{-\infty}^{t} e^{a(t-s)} P_2 f(s)ds - \int_{t}^{\infty} e^{a(t-s)} P_1 f(s)ds.$$

The problem we study here is to develop a similar theory for linear Volterra integral equations of the form

$$x(t) = f(t) + \int_0^t a(t-s)\, x(s)ds,$$

where $x, f \in R^n$ and a is an $n \times n$ matrix-valued function. In Section 2 we will show that the boundedness result cited above does have an analogue for the integral equation. For this aspect we shall assume that the "resolvent kernel" admits an exponential dichotomy. Then in Section 3 we shall seek sufficient conditions on the kernel $a(t)$ in order that the resolvent kernel admit an exponential dichotomy.

2. Resolvent Kernel with a Dichotomy

The integral equation

(2) $$x(t) = f(t) + \int_0^t a(t-s)x(s)ds, \quad -\infty < t < \infty,$$

can be solved by setting

(3) $$x(t) = f(t) - \int_0^t r(t-s)f(s)ds$$

where $r(t)$ is the matrix solution of

(4) $$r(t) = -a(t) + \int_0^t a(t-s)r(s)ds.$$

In these equations we view $x(t)$ and $f(t)$ as elements of the space $C = C(R, R^n)$ of continuous functions from R to R^n. The terms $a(t)$ and $r(t)$ are then $n \times n$ matrix valued functions. It is convenient to write Equations (2) and (3) in the abstract form

(2') $\qquad x = f + Ax$

and

(3') $\qquad x = f - Rf = (I - R)f,$

respectively.

Let us assume now that $r(t)$ admits an exponential dichotomy, that is, assume there exists projections P_1 and P_2 on R^n and positive constants K_1, K_2, m_1, m_2 such that $P_1 + P_2 = I$, $P_1P_2 = P_2P_1 = 0$ and

(5)
$$|r(t)P_1| \leq K_1 e^{m_1 t}, \quad t \leq 0$$
$$|r(t)P_2| \leq K_2 e^{-m_2 t}, \quad t \geq 0.$$

With P_1 and P_2 so given, we now define two operators L and B by

$$Lf(t) = \int_{-\infty}^{0} r(t-s)P_2 f(s)ds - \int_{0}^{\infty} r(t-s)\, P_1 f(s)ds$$

$$Bf(t) = \int_{-\infty}^{t} r(t-s)P_2 f(s)ds - \int_{t}^{\infty} r(t-s)\, P_1 f(s)ds.$$

It is then easily seen that Equation (3') can be rewritten as

$$x = f - Lf + Bf,$$

provided $f(t)$ is bounded on R, that is $f \in L_\infty \cap C$. Furthermore, we claim that $Bf \in L_\infty \cap C$ whenever $f \in L_\infty \cap C$. Indeed if we examine the first term in Bf we get

$$\left| \int_{-\infty}^{t} r(t-s)P_2 f(s)ds \right| = \left| \int_{0}^{\infty} r(s)P_2 f(t-s)ds \right| \leq \int_{0}^{\infty} |r(s)P_2|ds \cdot \|f\|_\infty$$
$$\leq K_2 m_2^{-1} \|f\|_\infty$$

where $\|f\|_\infty$ denotes the sup-norm of f. Similarly one has

$$\left| \int_t^{\infty} r(t-s)P_1 f(s)ds \right| \leq K_1 m_1^{-1} \|f\|_{\infty}.$$

Hence if f is bounded, $y = x + Lf$ is bounded. Furthermore y satisfies

(6) $$y = f^* + Ay$$

where

$$f^* = f + Bf - A(f + Bf) = (I - A)(I + B)f$$

Before summarizing these results, it is convenient to reformulate the results for ordinary differential equations which appear in Section 1. Recall that the general solution of Equation (1) can be written as

$$x(t) = e^{at}x_0 + \int_0^t e^{a(t-s)}f(s)ds.$$

We view $x(t)$ as the solution of Equation (1) with initial condition x_0.

If we form $y(t) = x(t) + z(t)$ where

$$z(t) = e^{at}\left[\int_{-\infty}^{0} e^{-as}P_2 f(s)ds - \int_0^{\infty} e^{-as}P_1 f(s)ds - x_0\right],$$

then $y(t)$ is the bounded solution of Equation (1) described in Section 1. In other words, given any solution x of Equation (1), there exists a function z (which is in fact a solution of the homogeneous equation $x' = ax$) such that $y = x + z$ is a bounded solution of Equation (1). Thus x and are solutions of the same differential equation, but satisfying, perhaps, different initial conditions.

Let us now return to the integral equation. In order to intepret the above results it is necessary to describe the phase space for the integral equation, that is, we must describe the space of initial data for the flow generated by the integral equation. However, by using the point of view described in [2] and [3], we see that the appropriate phase space for Equation (2) is $C = C(R, R^n)$.

In other words, the function f appearing in Equation (2) is the initial data and therefore Equation (5) represents the same equation (or flow) with a different initial condition. Thus we view x and y, the solutions of Equations (2) and (5), as solutions of the solutions of Equations (2) and (5), as solutions of the same flow but satisfying, perhaps, different initial conditions.

Let us now summarize this.

Theorem. Consider the linear integral equation

$$x(t) = f(t) + \int_0^t a(t - s)x(s)ds$$

where $f \in L_\infty \cap C$ and $a(t)$ is a locally-L_1-matrix-valued function. Assume that the resolvent kernel $r(t)$ given by Eqn. (4) admits an exponential dichotomy as described by (5). Then the solution y of

$$y(t) = f^*(t) + \int_0^t a(t-s)y(s)ds$$

is bounded where f^* is the initial condition

$$f^* = (I - A)\,(I + B)\,f.$$

3. Sufficient Conditions for a Dichotomy

The integral Equation (2) reduces to a differential equation when $a(t)$ is constant, say $a(t) \equiv a$. The solution $r(t)$ of the resolvent Equation (4) then becomes $r(t) = - ae^{at}$. As noted earlier, if the eigenvalues of a have nonzero real part, then $r(t)$ admits an exponential dichotomy.

One problem that deserves further study is to generalize the above observation to integral equations, that is to find sufficient conditions on $a(t)$ in order that the resolvent kernel $r(t)$ admit an exponential dichotomy. While we are unable to give a comprehensive answer to this problem at this time, we are able to state a result which in some ways generalizes the result state above for differential equations.

Assume that there exists a basis of vectors $\{x_1, \ldots, x_n\}$ in R^n satisfying

$$a(t)x_i = \gamma_i e^{\lambda_i t} x_i, \quad -\infty < t < \infty,$$

for appropriate complex numbers γ_i, λ_i. It follows then that

$$r(t)x_i = \gamma_i e^{(\lambda_i + \gamma_i)t} x_i, \quad -\infty < t < \infty.$$

Thus, if $\mathrm{Re}(\lambda_i + \gamma_i) \neq 0$ for all i, then r(t) admits an exponential dichotomy.

References

1. W. Coppel. Stability and Asymptotic Behavior of Differential Equations. Heath. Boston, 1965.

2. R. K. Miller and G. R. Sell. Volterra Integral Equations and Topological Dynamics. Amer. Math. Soc. Memoir No. 102, Providence, 1970.

3. G. R. Sell. "The Geometric Theory of Volterra Integral Equation. A Preliminary Report." Proceedings of Equadiff III Conference, Brno, 1972 (to appear).

An Entire Solution of the Functional Equation

$f(\lambda) + f(\omega\lambda)f(\omega^{-1}\lambda) = 1, \quad (\omega^5 = 1)$

Yasutaka Sibuya and Robert H. Cameron

1. Introduction

It has been proved by P.F. Hsieh and Y. Sibuya [2] that the linear differential equation

(1.1) $$d^2y/dx^2 - (x^3 + \lambda)\, y = 0$$

in which λ is a complex parameter has a unique solution

(1.2) $$y = \varphi(x, \lambda)$$

such that

(i) $\varphi(x, \lambda)$ is entire in (x, λ);

(ii) $\varphi(x, \lambda)$ admits an asymptotic representation

(1.3) $$\varphi(x, \lambda) = x^{-\frac{3}{4}} \left\{ 1 + 0(x^{-\frac{1}{2}}) \right\} \exp\left\{ -\frac{2}{5} x^{\frac{5}{2}} \right\}$$

uniformly on each compact set in the λ-plane as x tends to infinity in any closed subsector of the open sector

(1.4) $$\left| \arg x \right| < \frac{3\pi}{5} .$$

If we put

This research was supported in part by NSF Grants GP 27275 and GP 28732

(1.5) $$\omega = \exp\left\{i\frac{2\pi}{5}\right\}$$

we get five solutions

(1.6) $$y = \varphi_k(x,\lambda) = \varphi(\omega^{-k}x,\ \omega^{2k}\lambda)\ (k = 0,1,2,3,4)$$

of Equation (1.1). It is evident that we have

(1.7) $$\varphi(x,\lambda) = \varphi_0(x,\lambda) = \varphi_5(x,\lambda),$$

where

(1.8) $$\varphi_5(x,\lambda) = \varphi(\omega^{-5}x,\ \omega^{10}\lambda).$$

Let us denote by $\mathscr{S}_k$ a sector which is defined in the x-plane by

(1.9) $$\left|\arg x - k\frac{2\pi}{5}\right| < \frac{\pi}{5},$$

and we shall denote by $\bar{\mathscr{S}}_k$ the closure of $\mathscr{S}_k$. Then Sector (1.4) is given by $\mathscr{S}_{-1} \cup \bar{\mathscr{S}}_0 \cup \mathscr{S}_1$. Asymptotic property (ii) of $\varphi(x,\lambda)$ implies that

$$\varphi(x,\lambda) \to 0 \quad \text{as} \quad x \to \infty \quad \text{in } \mathscr{S}_0$$

and

$$\varphi(x,\lambda) \to \infty \quad \text{as} \quad x \to \infty \quad \text{in } \mathscr{S}_{-1} \text{ and } \mathscr{S}_1.$$

From this we conclude that

$$\varphi_k(x,\lambda) \to 0 \quad \text{as} \quad x \to \infty \quad \text{in } \mathscr{S}_k$$

and

$$\varphi_k(x,\lambda) \to \infty \quad \text{as} \quad x \to \infty \quad \text{in } \mathscr{S}_{k-1} \text{ and } \mathscr{S}_{k+1}.$$

Therefore φ_k and φ_{k+1} are linearly independent. In particular φ_1 and φ_2 are linearly independent. Hence φ_0 is a linear combination of φ_1 and φ_2. Y. Sibuya [3] proved that we have

(1.10) $$\varphi_0(x, \lambda) = c(\lambda)\, \varphi_1(x, \lambda) - \omega \varphi_2(x, \lambda)$$

where $c(\lambda)$ is an entire function of λ which admits asymptotic representations

(1.11) $$c(\lambda) = \omega^{\frac{1}{2}} \{1 + 0(1)\} \exp \left\{ K(1+ \omega^{-\frac{5}{6}})\, \lambda^{\frac{5}{6}} \right\}$$

as λ tendsto infinity in a sector

(1.12) $$-\frac{4\pi}{5} + \delta_0 \leqq \arg \lambda \leqq 2\pi - \frac{4\pi}{5} - \delta_0$$

and

(1.13) $$c(\lambda) = \omega^{\frac{1}{2}} \left[\{1 + 0(1)\} \exp \left\{ K(\omega^{-\frac{5}{6}} - \omega^{\frac{5}{6}})\, \lambda^{\frac{5}{6}} \right\} + \{1 + 0(1)\} \exp \left\{ K(1 + \omega^{-\frac{5}{6}})\, \lambda^{\frac{5}{6}} \right\} \right]$$

as λ tends to infinity in the sector

(1.14) $$\left| \arg \lambda + \frac{4\pi}{5} \right| \leqq \delta_0 \ ;$$

the positive constant δ_0 is arbitrary and the constant K is given by

$$K = \int_0^{+\infty} \left[(t^3 + 1)^{\frac{1}{2}} - t^{\frac{3}{2}} \right] dt > 0.$$

By using the identities

(1.15) $$\varphi_{k+1}(x, \lambda) = \varphi_k(\omega^{-1} x, \omega^2 \lambda)$$

we can derive from (1.10) the following relations:

(1.16-1) $$\varphi_1(x, \lambda) = c(\omega^2 \lambda)\, \varphi_2(x, \lambda) - \omega \varphi_3(x, \lambda) \quad ,$$

(1.16-2) $\varphi_2(x, \lambda) = c(\omega^4\lambda)\, \varphi_3(x, \lambda) - \omega\varphi_4(x, \lambda)$

and

(1.16-3) $\varphi_3(x, \lambda) = c(\omega^6\lambda)\, \varphi_4(x, \lambda) - \omega\varphi_5(x, \lambda)$.

Hence we obtain

(1.17) $$\omega^2\varphi_5(x, \lambda) = [1 - \omega^{-1}\, c(\omega\lambda)\, c(\omega^{-1}\lambda)]\ \varphi_1(x, \lambda)$$
$$+ [\omega^{-1}\, c(\omega\lambda)\, c(\omega^{-1}\lambda)\, c(\omega^2\lambda) - c(\omega\lambda) - c(\omega^2\lambda)]\, \varphi_2(x, \lambda)\ .$$

Finally from (1.7), (1.10) and (1.17) we derive

(1.18) $$\omega^2\, c(\lambda) = 1 - \omega^{-1}\, c(\omega\lambda)\, c(\omega^{-1}\lambda)$$

and

(1.19) $$-\omega^3 = \omega^{-1}\, c(\omega\lambda)\, c(\omega^{-1}\lambda)\, c(\omega^2\lambda) - c(\omega\lambda) - c(\omega^2\lambda).$$

It is easy to see that (1.18) implies (1.19). By putting

(1.20) $$f(\lambda) = \omega^2\, c(\lambda)$$

we can reduce (1.18) to

(1.21) $$f(\lambda) + f(\omega\lambda)\, f(\omega^{-1}\lambda) = 1.$$

We want to derive some information of the function $c(\lambda)$ from the relation (1.21). In this paper we shall present a result as the first step toward our goal.

The quantity $c(\lambda)$ is one of Stokes multipliers of Equation (1.1) at the irregular singular point at $x = \infty$. In deriving (1.21) we utilized the fact that the solution $\varphi(x, \lambda)$ is single-valued in the (x, λ)-space. In general, at an irregular singular point of a system of linear differential equations, we have some identities among solutions which are based on the so-called monodromy matrix. These identities induce various relations between Stokes multipliers. Relation (1.21) is one of the simplest examples of such relations.

2. Preliminary remarks on the function $f(\lambda)$

If we put $\lambda = 0$ in (1.21), we get

(2.1) $$f(0) + f(0)^2 = 1.$$

Two roots of this equation are $\omega + \omega^{-1}$ and $\omega^2 + \omega^{-2}$. We claim that

(2.2) $$f(0) = \omega^2 + \omega^{-2}.$$

In fact P.F. Hsieh and Y. Sibuya [2] showed that we have

(2.3) $$\varphi(x,0) = \gamma x^{\frac{1}{2}} H^{(1)}_{\frac{1}{5}}\left(\frac{2}{5} i x^{\frac{5}{2}}\right),$$

where $H^{(1)}_{\nu}$ is the Hankel function of the first kind of order ν, and γ is a suitable non-zero constant. From (1.6), (1.10) and (2.3) we derive

(2.4) $$H^{(1)}_{\frac{1}{5}}\left(\frac{2}{5} i x^{\frac{5}{2}}\right) = c(0)\,\omega^{-\frac{1}{2}} H^{(1)}_{\frac{1}{5}}\left(\frac{2}{5} i e^{-\pi i} x^{\frac{5}{2}}\right) - H^{(1)}_{\frac{1}{5}}\left(\frac{2}{5} i e^{-2\pi i} x^{\frac{5}{2}}\right).$$

On the other hand, it is known that

(2.5) $$\sin(\nu\pi)\, H^{(1)}_{\nu}(ze^{k\pi i}) = -\sin\{(k-1)\nu\pi\}\, H^{(1)}_{\nu}(z) - e^{-\nu\pi i}\sin(k\nu\pi)\, H^{(2)}_{\nu}(z)$$

and

(2.6) $$H^{(1)}_{\nu}(ze^{\pi i}) = -e^{-\nu\pi i} H^{(2)}_{\nu}(z),$$

where k is an integer. (See, for example, M. Abramowitz and L.A. Stegun [1; p.361, 9.1.38 and 9.1.39].) From those results, we obtain

(2.7) $$c(0) = 1 + \omega,$$

and hence we have (2.2).

The two constants $\omega + \omega^{-1}$ and $\omega^2 + \omega^{-2}$ are constant solutions of Equation (1.21). The function $f(\lambda)$ defined by (1.20) is a non-constant entire solution of Equation (1.21). It is very difficult to find a simple method to prove the existence of such an entire solution of Equation (1.21).

Let us write $f(\lambda)$ in a form

$$(2.8) \qquad f(\lambda) = G_0(z) + \lambda G_1(z) + \lambda^2 G_2(z) + \lambda^3 G_3(z) + \lambda^4 G_4(z),$$

where $G_j(z)$ are entire in z and $z = \lambda^5$. Since $f(0) = G_0(0)$, we have

$$(2.9) \qquad G_0(0) = \omega^2 + \omega^{-2} .$$

Put

$$(2.10) \qquad H(z) = 1 + (\omega^2 + \omega^{-2}) G_0(z) .$$

Then $H(z)$ is an entire function of z which is not identically equal to zero.

3. Main theorem

In this paper we shall prove the following theorem.

Theorem. The function $f(\lambda)$ has a representation

$$(3.1) \qquad f(\lambda) = \alpha \left\{ 1 + (1+\alpha^2) \frac{\xi(\lambda)\eta(\lambda) - \beta^2}{[\omega^2 \xi(\lambda) + \omega^{-2}\eta(\lambda) - 1][\omega^{-2}\xi(\lambda) + \omega^2 \eta(\lambda) - 1]} \right\}$$

where

$$(3.2) \qquad \alpha = \omega + \omega^{-1} , \quad \beta = \omega^2 + \omega^{-2}$$

and

$$(3.3) \qquad \xi(\lambda) = \lambda G_1(z)/H(z) , \quad \eta(\lambda) = \lambda^4 G_4(z)/H(z) .$$

Proof. Let

(3.4) $$f(\lambda) + f(\omega\lambda)\, f(\omega^{-1}\lambda) = X_0(z) + \lambda X_1(z) + \lambda^2 X_2(z) + \lambda^3 X_3(z) + \lambda^4 X_4(z),$$

where $X_j(z)$ are entire in z and $z = \lambda^5$. Then (1.21) implies $X_0(z) = 1$, $X_h(z) = 0$ $(h = 1,2,3,4)$. By computing X_0, X_1, X_2, X_3 and X_4 in terms of G_0, G_1, G_2, G_3 and G_4 we obtain

(3.5-0) $$G_0(z) + G_0(z)^2 + \beta zG_1(z)\, G_4(z) + \alpha\, zG_2(z)\, G_3(z) = 1,$$

(3.5-1) $$G_1(z) + \alpha G_0(z)\, G_1(z) + \beta zG_2(z)G_4(z) + zG_3(z)^2 = 0,$$

(3.5-2) $$G_2(z) + \beta G_0(z)\, G_2(z) + \alpha zG_3(z)\, G_4(z) + G_1(z)^2 = 0,$$

(3.5-3) $$G_3(z) + \beta G_0(z)G_3(z) + \alpha G_1(z)\, G_2(z) + zG_4(z)^2 = 0,$$

(3.5-4) $$G_4(z) + \alpha G_0(z)\, G_4(z) + \beta G_1(z)\, G_3(z) + G_2(z)^2 = 0.$$

From (3.5-2) and (3.5-3) we derive

(3.6) $$G_2(z) = \frac{\alpha z^2 G_4(z)^3 - H(z)G_1(z)^2}{H(z)^2 - \alpha^2 zG_1(z)G_4(z)} = \lambda^{-2}H(z)\,\frac{\alpha\,\eta(\lambda)^3 - \xi(\lambda)^2}{1 - \alpha^2\,\xi(\lambda)\,\eta(\lambda)}$$

and

(3.7) $$G_3(z) = \frac{\alpha G_1(z)^3 - zH(z)\, G_4(z)^2}{H(z)^2 - \alpha^2 zG_1(z)G_4(z)} = \lambda^{-3}H(z)\,\frac{\alpha\,\xi(\lambda)^3 - \eta(\lambda)^2}{1 - \alpha^2\,\xi(\lambda)\,\eta(\lambda)}.$$

If we insert (3.6) and (3.7) into (3.5-0) and if we use the identity

(3.8) $$1 - G_0(z) - G_0(z)^2 = (1 + \alpha G_0(z))(1 + \beta G_0(z)) = H(z)(1 + \alpha^2 - \alpha^2 H(z)),$$

we get

$$\frac{1+\alpha^2}{H(z)} - \alpha^2 = \beta\xi(\lambda)\eta(\lambda) + \alpha\,\frac{[\alpha\eta(\lambda)^3 - \xi(\lambda)^2][\alpha\xi(\lambda)^3 - \eta(\lambda)^2]}{[1-\alpha^2\xi(\lambda)\eta(\lambda)]^2}$$

and hence

$$(3.9)\qquad H(z) = \frac{(1+\alpha^2)[1-\alpha^2\xi(\lambda)\eta(\lambda)]^2}{[\alpha^2+\beta\xi(\lambda)\eta(\lambda)][1-\alpha^2\xi(\lambda)\eta(\lambda)]^2 + \alpha[\alpha\eta(\lambda)^3-\xi(\lambda)^2][\alpha\xi(\lambda)^3-\eta(\lambda)^2]}.$$

On the other hand, since $G_0 = \alpha(1-H)$, we have

$$f(\lambda) = \alpha - \alpha H(z) + H(z)[\xi(\lambda) + \eta(\lambda)]$$

$$+ H(z)\,\frac{\alpha[\xi(\lambda)^3 + \eta(\lambda)^3] - [\xi(\lambda)^2 + \eta(\lambda)^2]}{1-\alpha^2\xi(\lambda)\eta(\lambda)}$$

and hence

$$(3.10)\ f(\lambda) = \alpha\left\{1 - H(z)\,\frac{[1+\beta(\xi(\lambda)+\eta(\lambda))][1-\alpha^2\xi(\lambda)\eta(\lambda)] - [\xi(\lambda)^3+\eta(\lambda)^3] - \beta[\xi(\lambda)^2+\eta(\lambda)^2]}{1-\alpha^2\xi(\lambda)\eta(\lambda)}\right\}.$$

Note that

$$(3.11)\qquad [\alpha^2+\beta\xi\eta][1-\alpha^2\xi\eta]^2 + \alpha[\alpha\xi^3-\eta^2][\alpha\eta^3-\xi^2]$$

$$= \alpha^2[1-5\xi\eta+5(\xi\eta)^2-(\xi^5+\eta^5)]$$

$$= -\alpha^2(\xi+\eta-1)(\omega\xi+\omega^{-1}\eta-1)(\omega^{-1}\xi+\omega\eta-1)(\omega^2\xi+\omega^{-2}\eta-1)(\omega^{-2}\xi+\omega^2\eta-1)$$

and that

$$(3.12)\quad [1+\beta(\xi+\eta)][1-\alpha^2\xi\eta]-(\xi^3+\eta^3)-\beta(\xi^2+\eta^2)$$

$$= -(\xi+\eta-1)(\omega\xi+\omega^{-1}\eta-1)(\omega^{-1}\xi+\omega\eta-1).$$

We can easily derive (3.1) from (3.9), (3.10), (3.11) and (3.12). This completes the proof of our theorem.

In the proof of our theorem, we did not use (3.5-1) and (3.5-4). Actually, by using the fact that $H(z)$ is not identically equal to zero, we can prove that (3.5-1) and (3.5-4) are consequences of (3.5-0), (3.5-2) and (3.5-3).

References

1. M. Abramowitz and L.A. Stegun, Handbook of Mathematical Functions with Formulas, Graphs, and Mathematical Tables, U.S. Department of Commerce, National Bureau of Standards, Applied Mathematics Series, 55, 1964;

2. P. F. Hsieh and Y. Sibuya, "On the asymptotic integration of second order linear ordinary differential equations with polynomial coefficients", J. of Math. Anal. and Appl., 16(1966) 84-103;

3. Y. Sibuya, "Stokes multipliers of subdominant solutions of the differential equation $y'' - (x^3 + \lambda)y = 0$," Proc. Amer. Math. Soc., 18(1967) 238-243.

LIST OF CONTRIBUTORS

*Barnes, Earl R.

Mathematics Research Center, University of Wisconsin, Madison
Wisconsin 53706

*Cameron, Robert H.

School of Mathematics, University of Minnesota, Minneapolis, Minnesota
55455

Gollwitzer, Herman E.

Department of Mathematics, Drexel University, Philadelphia, Pennsylvania
19104

Grimm, Louis J.

Department of Mathematics, University of Missouri, Rolla
Rolla, Missouri 65401

*Hall, L. L.

Department of Mathematics, University of Missouri, Rolla
Rolla, Missouri 65401

Harris, William A.

Department of Mathematics, Univeristy of Southern California
Los Angeles, California 90007

Hsieh, Po-Fang

Department of Mathematics, Western Michigan University Kalamazoo
Michigan 49001

*Jurkat, Walfgang B.

Department of Mathematics, Syracuse University, Syracuse, New York 13224

Kimura, Tosihusa

School of Mathematics, University of Minnesota 55455
and Department of Mathematics, Faculty of Science, The University of
Tokyo, Tokyo, Japan

Loud, Warren S.

School of Mathematics, University of Minnesota, Minneapolis, Minnesota 55455

Lutz, Donald A.

Department of Mathematics, University of Wisconsin, Milwaukee Milwaukee, Wisconsin 53201

McGehee, Richard

School of Mathematics, University of Minnesota, Minneapolis, Minnesota 55455

Meyer, Kenneth

School of Mathematics, University of Minnesota, Minneapolis, Minnesota 55455

*Peyerimhoff, Alexander

Department of Mathematics, University of Ulm, Lindenstrasse 1, (7911) Unterelchingen, Fed. Rep. Germany

Rang, Edward R.

Corporate Research Center, Honeywell, Inc. Minneapolis, Minnesota

Russell, David L.

Department of Mathematics, University of Wisconsin, Madison, Wisconsin 53706

Sell, George R.

School of Mathematics, University of Minnesota, Minneapolis, Minnesota 55455

Sibuya, Yasutaka

School of Mathematics, University of Minnesota, Minneapolis, Minnesota 55455

Turrittin, Hugh L.

School of Mathematics, University of Minnesota, Minneapolis, Minnesota 55455

* Co-author of results presented by a participant.

Lecture Notes in Mathematics

Comprehensive leaflet on request

Vol. 146: A. B. Altman and S. Kleiman, Introduction to Grothendieck Duality Theory. II, 192 pages. 1970. DM 18,–

Vol. 147: D. E. Dobbs, Cech Cohomological Dimensions for Commutative Rings. VI, 176 pages. 1970. DM 16,–

Vol. 148: R. Azencott, Espaces de Poisson des Groupes Localement Compacts. IX, 141 pages. 1970. DM 16,–

Vol. 149: R. G. Swan and E. G. Evans, K-Theory of Finite Groups and Orders. IV, 237 pages. 1970. DM 20,–

Vol. 150: Heyer, Dualität lokalkompakter Gruppen. XIII, 372 Seiten. 1970. DM 20,–

Vol. 151: M. Demazure et A. Grothendieck, Schémas en Groupes I. (SGA 3). XV, 562 pages. 1970. DM 24,–

Vol. 152: M. Demazure et A. Grothendieck, Schémas en Groupes II. (SGA 3). IX, 654 pages. 1970. DM 24,–

Vol. 153: M. Demazure et A. Grothendieck, Schémas en Groupes III. (SGA 3). VIII, 529 pages. 1970. DM 24,–

Vol. 154: A. Lascoux et M. Berger, Variétés Kähleriennes Compactes. VII, 83 pages. 1970. DM 16,–

Vol. 155: Several Complex Variables I, Maryland 1970. Edited by J. Horváth. IV, 214 pages. 1970. DM 18,–

Vol. 156: R. Hartshorne, Ample Subvarieties of Algebraic Varieties. XIV, 256 pages. 1970. DM 20,–

Vol. 157: T. tom Dieck, K. H. Kamps und D. Puppe, Homotopietheorie. VI, 265 Seiten. 1970. DM 20,–

Vol. 158: T. G. Ostrom, Finite Translation Planes. IV. 112 pages. 1970. DM 16,–

Vol. 159: R. Ansorge und R. Hass. Konvergenz von Differenzenverfahren für lineare und nichtlineare Anfangswertaufgaben. VIII, 145 Seiten. 1970. DM 16,–

Vol. 160: L. Sucheston, Constributions to Ergodic Theory and Probability. VII, 277 pages. 1970. DM 20,–

Vol. 161: J. Stasheff, H-Spaces from a Homotopy Point of View. VI, 95 pages. 1970. DM 16,–

Vol. 162: Harish-Chandra and van Dijk, Harmonic Analysis on Reductive p-adic Groups. IV, 125 pages. 1970. DM 16,–

Vol. 163: P. Deligne, Equations Différentielles à Points Singuliers Reguliers. III, 133 pages. 1970. DM 16,–

Vol. 164: J. P. Ferrier, Seminaire sur les Algebres Complètes. II, 69 pages. 1970. DM 16,–

Vol. 165: J. M. Cohen, Stable Homotopy. V, 194 pages. 1970. DM 16,–

Vol. 166: A. J. Silberger, PGL_2 over the p-adics: its Representations, Spherical Functions, and Fourier Analysis. VII, 202 pages. 1970. DM 18,–

Vol. 167: Lavrentiev, Romanov and Vasiliev, Multidimensional Inverse Problems for Differential Equations. V, 59 pages. 1970. DM 16,–

Vol. 168: F. P. Peterson, The Steenrod Algebra and its Applications: A conference to Celebrate N. E. Steenrod's Sixtieth Birthday. VII, 317 pages. 1970. DM 22,–

Vol. 169: M. Raynaud, Anneaux Locaux Henséliens. V, 129 pages. 1970. DM 16,–

Vol. 170: Lectures in Modern Analysis and Applications III. Edited by C. T. Taam. VI, 213 pages. 1970. DM 18,–

Vol. 171: Set-Valued Mappings, Selections and Topological Properties of 2^X. Edited by W. M. Fleischman. X, 110 pages. 1970. DM 16,–

Vol. 172: Y.-T. Siu and G. Trautmann, Gap-Sheaves and Extension of Coherent Analytic Subsheaves. V, 172 pages. 1971. DM 16,–

Vol. 173: J. N. Mordeson and B. Vinograde, Structure of Arbitrary Purely Inseparable Extension Fields. IV, 138 pages. 1970. DM 16,–

Vol. 174: B. Iversen, Linear Determinants with Applications to the Picard Scheme of a Family of Algebraic Curves. VI, 69 pages. 1970. DM 16,–

Vol. 175: M. Brelot, On Topologies and Boundaries in Potential Theory. VI, 176 pages. 1971. DM 18,–

Vol. 176: H. Popp, Fundamentalgruppen algebraischer Mannigfaltigkeiten. IV, 154 Seiten. 1970. DM 16,–

Vol. 177: J. Lambek, Torsion Theories, Additive Semantics and Rings of Quotients. VI, 94 pages. 1971. DM 16,–

Vol. 178: Th. Bröcker und T. tom Dieck, Kobordismentheorie. XVI, 191 Seiten. 1970. DM 18,–

Vol. 179: Seminaire Bourbaki – vol. 1968/69. Exposés 347-363. IV. 295 pages. 1971. DM 22,–

Vol. 180: Séminaire Bourbaki – vol. 1969/70. Exposés 364-381. IV, 310 pages. 1971. DM 22,–

Vol. 181: F. DeMeyer and E. Ingraham, Separable Algebras over Commutative Rings. V, 157 pages. 1971. DM 16.–

Vol. 182: L. D. Baumert. Cyclic Difference Sets. VI, 166 pages. 1971. DM 16,–

Vol. 183: Analytic Theory of Differential Equations. Edited by P. F. Hsieh and A. W. J. Stoddart. VI, 225 pages. 1971. DM 20,–

Vol. 184: Symposium on Several Complex Variables, Park City, Utah, 1970. Edited by R. M. Brooks. V, 234 pages. 1971. DM 20,–

Vol. 185: Several Complex Variables II, Maryland 1970. Edited by J. Horváth. III, 287 pages. 1971. DM 24,–

Vol. 186: Recent Trends in Graph Theory. Edited by M. Capobianco/ J. B. Frechen/M. Krolik. VI, 219 pages. 1971. DM 18.–

Vol. 187: H. S. Shapiro, Topics in Approximation Theory. VIII, 275 pages. 1971. DM 22,–

Vol. 188: Symposium on Semantics of Algorithmic Languages. Edited by E. Engeler. VI, 372 pages. 1971. DM 26,–

Vol. 189: A. Weil, Dirichlet Series and Automorphic Forms. V, 164 pages. 1971. DM 16,–

Vol. 190: Martingales. A Report on a Meeting at Oberwolfach, May 17-23, 1970. Edited by H. Dinges. V, 75 pages. 1971. DM 16,–

Vol. 191: Séminaire de Probabilités V. Edited by P. A. Meyer. IV, 372 pages. 1971. DM 26,–

Vol. 192: Proceedings of Liverpool Singularities – Symposium I. Edited by C. T. C. Wall. V, 319 pages. 1971. DM 24,–

Vol. 193: Symposium on the Theory of Numerical Analysis. Edited by J. Ll. Morris. VI, 152 pages. 1971. DM 16,–

Vol. 194: M. Berger, P. Gauduchon et E. Mazet. Le Spectre d'une Variété Riemannienne. VII, 251 pages. 1971. DM 22,–

Vol. 195: Reports of the Midwest Category Seminar V. Edited by J.W. Gray and S. Mac Lane. III, 255 pages. 1971. DM 22,–

Vol. 196: H-spaces – Neuchâtel (Suisse)- Août 1970. Edited by F. Sigrist, V, 156 pages. 1971. DM 16,–

Vol. 197: Manifolds – Amsterdam 1970. Edited by N. H. Kuiper. V, 231 pages. 1971. DM 20,–

Vol. 198: M. Hervé, Analytic and Plurisubharmonic Functions in Finite and Infinite Dimensional Spaces. VI, 90 pages. 1971. DM 16.–

Vol. 199: Ch. J. Mozzochi, On the Pointwise Convergence of Fourier Series. VII, 87 pages. 1971. DM 16,–

Vol. 200: U. Neri, Singular Integrals. VII, 272 pages. 1971. DM 22,–

Vol. 201: J. H. van Lint, Coding Theory. VII, 136 pages. 1971. DM 16,–

Vol. 202: J. Benedetto, Harmonic Analysis on Totally Disconnected Sets. VIII, 261 pages. 1971. DM 22,–

Vol. 203: D. Knutson, Algebraic Spaces. VI, 261 pages. 1971. DM 22,–

Vol. 204: A. Zygmund, Intégrales Singulières. IV, 53 pages. 1971. DM 16,–

Vol. 205: Séminaire Pierre Lelong (Analyse) Année 1970. VI, 243 pages. 1971. DM 20,–

Vol. 206: Symposium on Differential Equations and Dynamical Systems. Edited by D. Chillingworth. XI, 173 pages. 1971. DM 16,–

Vol. 207: L. Bernstein, The Jacobi-Perron Algorithm – Its Theory and Application. IV, 161 pages. 1971. DM 16,–

Vol. 208: A. Grothendieck and J. P. Murre, The Tame Fundamental Group of a Formal Neighbourhood of a Divisor with Normal Crossings on a Scheme. VIII, 133 pages. 1971. DM 16,–

Vol. 209: Proceedings of Liverpool Singularities Symposium II. Edited by C. T. C. Wall. V, 280 pages. 1971. DM 22,–

Vol. 210: M. Eichler, Projective Varieties and Modular Forms. III, 118 pages. 1971. DM 16,–

Vol. 211: Théorie des Matroïdes. Edité par C. P. Bruter. III, 108 pages. 1971. DM 16,–

Please turn over

Vol. 212: B. Scarpellini, Proof Theory and Intuitionistic Systems. VII, 291 pages. 1971. DM 24,–

Vol. 213: H. Hogbe-Nlend, Théorie des Bornologies et Applications. V, 168 pages. 1971. DM 18,–

Vol. 214: M. Smorodinsky, Ergodic Theory, Entropy. V, 64 pages. 1971. DM 16,–

Vol. 215: P. Antonelli, D. Burghelea and P. J. Kahn, The Concordance-Homotopy Groups of Geometric Automorphism Groups. X, 140 pages. 1971. DM 16,–

Vol. 216: H. Maaß, Siegel's Modular Forms and Dirichlet Series. VII, 328 pages. 1971. DM 20,–

Vol. 217: T. J. Jech, Lectures in Set Theory with Particular Emphasis on the Method of Forcing. V, 137 pages. 1971. DM 16,–

Vol. 218: C. P. Schnorr, Zufälligkeit und Wahrscheinlichkeit. IV, 212 Seiten 1971. DM 20,–

Vol. 219: N. L. Alling and N. Greenleaf, Foundations of the Theory of Klein Surfaces. IX, 117 pages. 1971. DM 16,–

Vol. 220: W. A. Coppel, Disconjugacy. V, 148 pages. 1971. DM 16,–

Vol. 221: P. Gabriel und F. Ulmer, Lokal präsentierbare Kategorien. V, 200 Seiten. 1971. DM 18,–

Vol. 222: C. Meghea, Compactification des Espaces Harmoniques. III, 108 pages. 1971. DM 16,–

Vol. 223: U. Felgner, Models of ZF-Set Theory. VI, 173 pages. 1971. DM 16,–

Vol. 224: Revètements Etales et Groupe Fondamental. (SGA 1). Dirigé par A. Grothendieck XXII, 447 pages. 1971. DM 30,–

Vol. 225: Théorie des Intersections et Théorème de Riemann-Roch. (SGA 6). Dirigé par P. Berthelot, A. Grothendieck et L. Illusie. XII, 700 pages. 1971. DM 40,–

Vol. 226: Seminar on Potential Theory, II. Edited by H. Bauer. IV, 170 pages. 1971. DM 18,–

Vol. 227: H. L. Montgomery, Topics in Multiplicative Number Theory. IX, 178 pages. 1971. DM 18,–

Vol. 228: Conference on Applications of Numerical Analysis. Edited by J. Ll. Morris. X, 358 pages. 1971. DM 26,–

Vol. 229: J. Väisälä, Lectures on n-Dimensional Quasiconformal Mappings. XIV, 144 pages. 1971. DM 16,–

Vol. 230: L. Waelbroeck, Topological Vector Spaces and Algebras. VII, 158 pages. 1971. DM 16,–

Vol. 231: H. Reiter, L^1-Algebras and Segal Algebras. XI, 113 pages. 1971. DM 16,–

Vol. 232: T. H. Ganelius, Tauberian Remainder Theorems. VI, 75 pages. 1971. DM 16,–

Vol. 233: C. P. Tsokos and W. J. Padgett. Random Integral Equations with Applications to Stochastic Systems. VII, 174 pages. 1971. DM 18,–

Vol. 234: A. Andreotti and W. Stoll. Analytic and Algebraic Dependence of Meromorphic Functions. III, 390 pages. 1971. DM 26,–

Vol. 235: Global Differentiable Dynamics. Edited by O. Hájek, A. J. Lohwater, and R. McCann. X, 140 pages. 1971. DM 16,–

Vol. 236: M. Barr, P. A. Grillet, and D. H. van Osdol. Exact Categories and Categories of Sheaves. VII, 239 pages. 1971, DM 20,–

Vol. 237: B. Stenström. Rings and Modules of Quotients. VII, 136 pages. 1971. DM 16,–

Vol. 238: Der kanonische Modul eines Cohen-Macaulay-Rings. Herausgegeben von Jürgen Herzog und Ernst Kunz. VI, 103 Seiten. 1971. DM 16,–

Vol. 239: L. Illusie, Complexe Cotangent et Déformations I. XV, 355 pages. 1971. DM 26,–

Vol. 240: A. Kerber, Representations of Permutation Groups I. VII, 192 pages. 1971. DM 18,–

Vol. 241: S. Kaneyuki, Homogeneous Bounded Domains and Siegel Domains. V, 89 pages. 1971. DM 16,–

Vol. 242: R. R. Coifman et G. Weiss, Analyse Harmonique Non-Commutative sur Certains Espaces. V, 160 pages. 1971. DM 16,–

Vol. 243: Japan-United States Seminar on Ordinary Differential and Functional Equations. Edited by M. Urabe. VIII, 332 pages. 1971. DM 26,–

Vol. 244: Séminaire Bourbaki – vol. 1970/71. Exposés 382–399. IV, 356 pages. 1971. DM 26,–

Vol. 245: D. E. Cohen, Groups of Cohomological Dimension One. V, 99 pages. 1972. DM 16,–

Vol. 246: Lectures on Rings and Modules. Tulane University Ring and Operator Theory Year, 1970–1971. Volume I. X, 661 pages. 1972. DM 40,–

Vol. 247: Lectures on Operator Algebras. Tulane University Ring and Operator Theory Year, 1970–1971. Volume II. XI, 786 pages. 1972. DM 40,–

Vol. 248: Lectures on the Applications of Sheaves to Ring Theory. Tulane University Ring and Operator Theory Year, 1970–1971. Volume III. VIII, 315 pages. 1971. DM 26,–

Vol. 249: Symposium on Algebraic Topology. Edited by P. J. Hilton. VII, 111 pages. 1971. DM 16,–

Vol. 250: B. Jónsson, Topics in Universal Algebra. VI, 220 pages. 1972. DM 20,–

Vol. 251: The Theory of Arithmetic Functions. Edited by A. A. Gioia and D. L. Goldsmith VI, 287 pages. 1972. DM 24,–

Vol. 252: D. A. Stone, Stratified Polyhedra. IX, 193 pages. 1972. DM 18,–

Vol. 253: V. Komkov, Optimal Control Theory for the Damping of Vibrations of Simple Elastic Systems. V, 240 pages. 1972. DM 20,–

Vol. 254: C. U. Jensen, Les Foncteurs Dérivés de $\varprojlim$ et leurs Applications en Théorie des Modules. V, 103 pages. 1972. DM 16,–

Vol. 255: Conference in Mathematical Logic – London '70. Edited by W. Hodges. VIII, 351 pages. 1972. DM 26,–

Vol. 256: C. A. Berenstein and M. A. Dostal, Analytically Uniform Spaces and their Applications to Convolution Equations. VII, 130 pages. 1972. DM 16,–

Vol. 257: R. B. Holmes, A Course on Optimization and Best Approximation. VIII, 233 pages. 1972. DM 20,–

Vol. 258: Séminaire de Probabilités VI. Edited by P. A. Meyer. VI, 253 pages. 1972. DM 22,–

Vol. 259: N. Moulis, Structures de Fredholm sur les Variétés Hilbertiennes. V, 123 pages. 1972. DM 16,–

Vol. 260: R. Godement and H. Jacquet, Zeta Functions of Simple Algebras. IX, 188 pages. 1972. DM 18,–

Vol. 261: A. Guichardet, Symmetric Hilbert Spaces and Related Topics. V, 197 pages. 1972. DM 18,–

Vol. 262: H. G. Zimmer, Computational Problems, Methods, and Results in Algebraic Number Theory. V, 103 pages. 1972. DM 16,–

Vol. 263: T. Parthasarathy, Selection Theorems and their Applications. VII, 101 pages. 1972. DM 16,–

Vol. 264: W. Messing, The Crystals Associated to Barsotti-Tate Groups: with Applications to Abelian Schemes. III, 190 pages. 1972. DM 18,–

Vol. 265: N. Saavedra Rivano, Catégories Tannakiennes. II, 418 pages. 1972. DM 26,–

Vol. 266: Conference on Harmonic Analysis. Edited by D. Gulick and R. L. Lipsman. VI, 323 pages. 1972. DM 24,–

Vol. 267: Numerische Lösung nichtlinearer partieller Differential- und Integro-Differentialgleichungen. Herausgegeben von R. Ansorge und W. Törnig, VI, 339 Seiten. 1972. DM 26,–

Vol. 268: C. G. Simader, On Dirichlet's Boundary Value Problem. IV, 238 pages. 1972. DM 20,–

Vol. 269: Théorie des Topos et Cohomologie Etale des Schémas. (SGA 4). Dirigé par M. Artin, A. Grothendieck et J. L. Verdier. XIX, 525 pages. 1972. DM 50,–

Vol. 270: Thèorie des Topos et Cohomologie Etle des Schémas. Tome 2. (SGA 4). Dirige par M. Artin, A. Grothendieck et J. L. Verdier. V, 418 pages. 1972. DM 50,–

Vol. 271: J. P. May, The Geometry of Iterated Loop Spaces. IX, 175 pages. 1972. DM 18,–

Vol. 272: K. R. Parthasarathy and K. Schmidt, Positive Definite Kernels, Continuous Tensor Products, and Central Limit Theorems of Probability Theory. VI, 107 pages. 1972. DM 16,–

Vol. 273: U. Seip, Kompakt erzeugte Vektorräume und Analysis. IX, 119 Seiten. 1972. DM 16,–

Vol. 274: Toposes, Algebraic Geometry and Logic. Edited by. F. W. Lawvere. VI, 189 pages. 1972. DM 18,–

Vol. 275: Séminaire Pierre Lelong (Analyse) Année 1970–1971. VI, 181 pages. 1972. DM 18,–

Vol. 276: A. Borel, Représentations de Groupes Localement Compacts. V, 98 pages. 1972. DM 16,–

Vol. 277: Séminaire Banach. Edité par C. Houzel. VII, 229 pages. 1972. DM 20,–